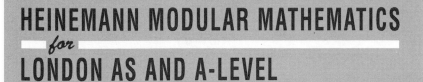

HEINEMANN MODULAR MATHEMATICS
for
LONDON AS AND A-LEVEL
Mechanics 2

Jean Littlewood   John Hebborn

Heinemann

Heinemann Educational Publishers,
a division of Heinemann Publishers (Oxford) Ltd,
Halley Court, Jordan Hill, Oxford, OX2 8EJ

OXFORD MELBOURNE AUCKLAND
JOHANNESBURG BLANTYRE GABORONE
IBADAN PORTSMOUTH (NH) USA CHICAGO

First published 1995

02 01 00

10 9 8

ISBN 0 435 51804 6

Original design by Geoffrey Wadsley: additional design work by Jim Turner

Typeset and illustrated by TecSet Limited, Wallington, Surrey

Printed in Great Britain by The Bath Press, Bath

Acknowledgements:

The publisher's and authors' thanks are due to the University of London
Examinations and Assessment Council (ULEAC) for permission to reproduce
questions from past examination papers. These are marked with an [L].
    The answers have been provided by the authors and are not the responsibility
of the examining board.

About this book

This book is designed to provide you with the best preparation possible for your London Modular Mathematics M2 examination. The series authors are examiners and exam moderators themselves and have a good understanding of the exam board's requirements.

Finding your way around

To help to find your way around when you are studying and revising use the:

- **edge marks** (shown on the front page) – these help you to get to the right chapter quickly;
- **contents list** – this lists the headings that identify key syllabus ideas covered in the book so you can turn straight to them;
- **index** – if you need to find a topic the **bold** number shows where to find the main entry on a topic.

Remembering key ideas

We have provided clear explanations of the key ideas and techniques you need throughout the book. Key ideas you need to remember are listed in a **summary of key points** at the end of each chapter and marked like this in the chapters:

$$\blacksquare \qquad \textbf{work done} = \int_{x_1}^{x_2} \mathrm{F}(x)\,\mathrm{d}x$$

Exercises and exam questions

In this book questions are carefully graded so they increase in difficulty and gradually bring you up to exam standard.

- **past exam questions** are marked with an L;
- **review exercises** on pages 99 and 225 help you practise answering questions from several areas of mathematics at once, as in the real exam;
- **exam style practice paper** – this is designed to help you prepare for the exam itself;
- **answers** are included at the end of the book – use them to check your work.

Contents

A knowledge and expertise in the contents of Book M1 and chapters 8 and 9 (differentiation and integration) of Book P2 is assumed and expected.

Further kinematics

Chapter 3 of book M1 dealt with the kinematics of a particle when the acceleration is constant. Remember that the constant acceleration formulae derived in section 3.1 of Book M1 must only be used when the acceleration is constant. However, in practice, the acceleration of a particle is often not constant. This chapter revises and extends the material covered in section 3.4 of Book M1 and considers the kinematics of a particle moving in one dimension, when the acceleration is not constant.

Two cases will be considered:
(1) when the acceleration depends on time,
(2) when the acceleration depends on the displacement.

In both cases calculus methods must be used.

1.1 Acceleration depending on time (*t*)

Consider a particle P moving in a straight line. Suppose that at time t seconds the displacement of P from a fixed point O in the line is x metres, its velocity is $v\,\mathrm{m\,s^{-1}}$ and its acceleration is $a\,\mathrm{m\,s^{-2}}$.

You will recall from chapter 3 of book M1 that: $a = \dfrac{\mathrm{d}v}{\mathrm{d}t}$

and: $v = \dfrac{\mathrm{d}x}{\mathrm{d}t}$

so: $a = \dfrac{\mathrm{d}^2 x}{\mathrm{d}t^2}$

When the acceleration is a function of time:

$$a = \mathrm{f}(t)$$

So: $v = \displaystyle\int \mathrm{f}(t)\,\mathrm{d}t + c$

where c is an arbitrary constant of integration.

Similarly, when $v = g(t)$:

then:
$$x = \int g(t)dt + k$$

where k is an arbitrary constant of integration.

Example 1

A particle P is moving along the line Ox with a constant acceleration equal to $A\,\mathrm{m\,s^{-2}}$ $(A > 0)$. When $t = 0$ the particle is at O, so that $x = 0$, and it has a speed of $u\,\mathrm{m\,s^{-1}}$ in the direction Ox.

(a) Obtain, by using calculus, v and x as functions of t.
(b) Sketch the graphs of v against t and x against t.

(a) As:
$$a = \frac{dv}{dt} = A$$

So:
$$v = \int A dt + c$$
$$= At + c, \text{ since } A \text{ is constant.}$$

The value of the constant c may be found by using $v = u$ when $t = 0$.

Then:
$$u = 0 + c$$

$\Rightarrow$
$$c = u$$

and:
$$v = u + At$$

As:
$$v = \frac{dx}{dt} = u + At$$

so:
$$x = \int (u + At)dt + k$$
$$= ut + \tfrac{1}{2}At^2 + k, \text{ since both } A \text{ and } u \text{ are constant.}$$

As $x = 0$ when $t = 0$ the constant k may be found.

Since:
$$0 = 0 + 0 + k$$

then:
$$k = 0$$

and so:
$$x = ut + \tfrac{1}{2}At^2$$

(b) The sketch of v against t is a straight line with gradient A which cuts the v-axis at $v = u$.

To sketch x against t note:

(1) The gradient of the curve is $(u + At)$ and this is always positive for $t > 0$, as both u and A are positive. The gradient is zero when $t = -\dfrac{u}{A}$.

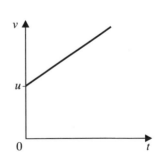

(2) $x = 0$ when $ut + \frac{1}{2}At^2 = 0$

That is, when: $\qquad\qquad t(u + \frac{1}{2}At) = 0$

$\Rightarrow \qquad\qquad\qquad t = 0$ or $t = -\dfrac{2u}{A}$

So for $t \geqslant 0$ the only solution is $t = 0$.

(3) When $t = 0$, $x = 0$ so the curve goes through (0,0). The sketch of x against t may now be drawn.

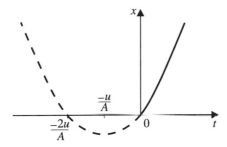

Example 2

A particle P is moving along the line Ox with an acceleration $\frac{2}{3}t\,\mathrm{m\,s^{-2}}$ at time t seconds. When $t = 3$ the speed of P is $4\,\mathrm{m\,s^{-1}}$ in the direction Ox and its displacement from O is $10\,\mathrm{m}$. Find the displacement of P from O when $t = 6$.

As: $\qquad\qquad a = \dfrac{\mathrm{d}v}{\mathrm{d}t} = \dfrac{2}{3}t$

so: $\qquad\qquad v = \displaystyle\int \dfrac{2}{3}t \, \mathrm{d}t + c$

$$= \frac{2}{3}\left(\frac{t^2}{2}\right) + c = \frac{1}{3}t^2 + c$$

where c is an arbitrary constant.

Since $v = 4$ when $t = 3$

$$4 = \frac{1}{3}(3)^2 + c$$

So: $\qquad\qquad c = 1$

and: $\qquad\qquad v = \dfrac{1}{3}t^2 + 1$

From: $\qquad\qquad v = \dfrac{\mathrm{d}x}{\mathrm{d}t} = \dfrac{1}{3}t^2 + 1$

integrating with respect to t gives:

$$x = \int \left(\frac{1}{3}t^2 + 1\right) dt + k$$

$$= \frac{1}{3} \times \frac{t^3}{3} + t + k$$

$$= \frac{1}{9}t^3 + t + k$$

where k is an arbitrary constant.

Since $x = 10$ when $t = 3$

$$10 = \frac{1}{9}(3)^3 + 3 + k$$

So: $\qquad\qquad\qquad k = 4$

and: $\qquad\qquad\qquad x = \frac{1}{9}t^3 + t + 4$

When $t = 6$: $\qquad\qquad x = \frac{1}{9}(6)^3 + 6 + 4 = 34$

The displacement of P from O when $t = 6$ is 34 m.

Example 3

A particle P is moving along the line Ox with an acceleration $(6t - 18)\,\text{m s}^{-2}$ at time t seconds. When $t = 0$ the displacement of P from O is zero and its speed is $24\,\text{m s}^{-1}$ in the direction Ox.

(a) Find the values of t for which $v = 0$ and the values of x for these two values of t.
(b) Find the distance travelled in the first 3 seconds.

On first sight this question may appear to be exactly the same as example 2. There are however some important differences. In example 2 the acceleration was only zero when $t = 0$ and was positive for all $t > 0$. The speed $v\,\text{m s}^{-1}$ never vanished – it was in fact always positive and so x increased with t.

In the solution the differences between this example and example 2 will be highlighted.

(a) As: $\qquad\qquad a = \dfrac{dv}{dt} = 6t - 18$

integrating with respect to t gives:

$$v = \int (6t - 18)\mathrm{d}t + c$$

$$= 6 \times \frac{t^2}{2} - 18t + c$$

$$= 3t^2 - 18t + c$$

where c is an arbitrary constant.

Since $v = 24$ when $t = 0$

$$24 = 0 - 0 + c$$

So: $\qquad\qquad c = 24$

and: $\qquad\qquad v = 3t^2 - 18t + 24$ $\qquad\qquad$ (1)

The values of t for which $v = 0$ are obtained from

$$3t^2 - 18t + 24 = 0$$

or: $\qquad\qquad t^2 - 6t + 8 = 0$

Factorising: $\qquad (t - 4)(t - 2) = 0$

so: $\qquad\qquad t = 2 \;\; \text{or} \;\; t = 4$

It is useful to sketch v against t for this situation as this gives us a clear picture of how v varies with time. The following facts will enable us to produce this sketch.

(1) $v = 0$ when $t = 2$ and $t = 4$

(2) when $t = 0$, $v = 24$

(3) v has a stationary value when $\dfrac{\mathrm{d}v}{\mathrm{d}t}$, that is a, is zero (see section 8.8 of Book P1). So this occurs when $6t - 18 = 0$, that is when $t = 3$.

The sketch is:

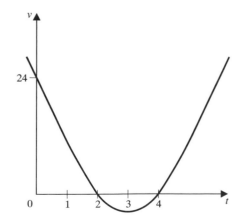

To find the distances travelled by P when $t = 2$ and $t = 4$ use:

$$v = \frac{\mathrm{d}x}{\mathrm{d}t} = 3t^2 - 18t + 24 \text{ from equation (1).}$$

Integrating with respect to t:

$$x = 3 \times \frac{t^3}{3} - 18 \times \frac{t^2}{2} + 24t + k$$
$$= t^3 - 9t^2 + 24t + k$$

where k is an arbitrary constant.

Since $x = 0$ when $t = 0$

$$0 = 0 - 0 + 0 + k$$

So: $\qquad\qquad\qquad k = 0$

and: $\qquad\qquad\qquad x = t^3 - 9t^2 + 24t$ $\qquad\qquad$ (2)

When $t = 2$ the value of x is given by:

$$x = (2)^3 - 9(2)^2 + 24(2)$$

So $x = 20\,\mathrm{m}$ and P is at a point A.

When $t = 4$ the value of x is given by:

$$x = (4)^3 - 9(4)^2 + 24(4)$$

So $x = 16\,\mathrm{m}$ and P is at a point B.

These solutions are illustrated below on the line Ox together with the initial situation.

```
t = 0                          t = 4        t = 2
├───────────────────────────────┼────────────┤
O                               B            A
                              16 m         20 m
```

Notice that the displacement of P from O when $t = 4$ is smaller than the displacement of P from O when $t = 2$. This is because v is negative for $2 < t < 4$, as shown in the sketch of v against t, and since $v = \frac{\mathrm{d}x}{\mathrm{d}t}$, x must be decreasing for $2 < t < 4$.

(b) The displacement of P from O when $t = 3$ is obtained by substituting $t = 3$ into equation (2).

Hence: $\qquad\qquad\qquad x = (3)^3 - 9(3)^2 + 24(3)$

$$= 18$$

So $x = 18\,\mathrm{m}$ and P is at a point C between A and B.

In the first two seconds v is positive and P travels from O to A a distance of 20 m. In the third second v is negative so the direction of motion is reversed and P moves from A to C a distance of $(20 - 18)\,\text{m} = 2\,\text{m}$.

The total distance travelled in the first 3 seconds is therefore

$$OA + AC = 20\,\text{m} + 2\,\text{m}$$
$$= 22\,\text{m}$$

Example 4

A particle P starts from rest at O at time $t = 0$ and moves with variable acceleration $a\ \text{m s}^{-2}$ where

$$a = \frac{t}{5}, \qquad 0 \leqslant t < 5$$
$$a = \frac{t}{5} + \frac{10}{t^2}, \qquad t \geqslant 5$$

where t is measured in seconds. Show that the speed of P is $2\frac{1}{2}\,\text{m s}^{-1}$ when $t = 5$ and $11\,\text{m s}^{-1}$ when $t = 10$. Show that the displacement of P from O when $t = 10$ is $(43\frac{1}{3} - 10\ln 2)$ metres.

As a has a different form in the two regions these regions must be considered separately. The fact that v and x must be continuous at $t = 5$, the boundary between the two regions, must then be used.

For $0 \leqslant t < 5$ (region I)

$$a = \frac{\mathrm{d}v}{\mathrm{d}t} = \frac{t}{5}$$

So, integrating with respect to t:

$$v = \int \frac{t}{5}\,\mathrm{d}t = \frac{1}{5}\left(\frac{t^2}{2}\right) + c = \frac{t^2}{10} + c$$

where c is an arbitrary constant. The value of c can be found since P starts from rest at $t = 0$.

Hence: $\qquad\qquad v = 0$ at $t = 0 \Rightarrow 0 = 0 + c$

So: $\qquad\qquad\qquad c = 0$

and: $\qquad\qquad\quad v = \frac{t^2}{10}, \ 0 \leqslant t < 5$ $\hfill$ (1)

Hence when $t = 5$: $\qquad v = \dfrac{5^2}{10} = 2\tfrac{1}{2}$

The speed of P is $2\tfrac{1}{2}\,\mathrm{m\,s^{-1}}$ when $t = 5$.

For $t \geqslant 5$ (region II)

$$a = \frac{\mathrm{d}v}{\mathrm{d}t} = \left(\frac{t}{5} + \frac{10}{t^2}\right)$$

So, integrating with respect to t:

$$v = \int\left(\frac{t}{5} + \frac{10}{t^2}\right)\mathrm{d}t = \frac{t^2}{10} + \frac{10t^{-1}}{(-1)} + k$$

$$= \frac{t^2}{10} - \frac{10}{t} + k$$

where k is an arbitrary constant.

To find k use the result obtained above, that is $v = 2\tfrac{1}{2}$ when $t = 5$.

So: $\qquad 2\tfrac{1}{2} = \dfrac{25}{10} - \dfrac{10}{5} + k$

and: $\qquad k = 2$

Hence: $\qquad v = \dfrac{t^2}{10} - \dfrac{10}{t} + 2,\ t \geqslant 5$ $\qquad\qquad$ (2)

When $t = 10$ we have:

$$v = \frac{100}{10} - \frac{10}{10} + 2 = 11$$

The speed of P is $11\,\mathrm{m\,s^{-1}}$ when $t = 10$.

To find the displacement of P you must return to equation (1). In region I for $0 \leqslant t < 5$:

$$v = \frac{\mathrm{d}x}{\mathrm{d}t} = \frac{t^2}{10}$$

So, integrating with respect to t gives:

$$x = \frac{1}{10}\left(\frac{t^3}{3}\right) + A$$

where A is an arbitrary constant. Taking $x\,$m to be the displacement from O and using the condition $x = 0$ at $t = 0$, since P starts from O, gives:

$$0 = 0 + A$$

So: $\qquad A = 0$

and: $\qquad x = \dfrac{t^3}{30},\ 0 \leqslant t < 5$

When $t = 5$ you find that $x = \dfrac{5^3}{30} = 4\frac{1}{6}$ m.

In region II, using equation (2), for $t \geqslant 5$

$$v = \frac{\mathrm{d}x}{\mathrm{d}t} = \frac{t^2}{10} - \frac{10}{t} + 2$$

So, integrating with respect to t gives:

$$x = \frac{1}{10}\left(\frac{t^3}{3}\right) - 10\ln t + 2t + B$$

$$= \frac{t^3}{30} - 10\ln t + 2t + B$$

where B is an arbitrary constant.

From above, $x = 4\frac{1}{6}$ when $t = 5$ and substituting these values gives:

$$4\frac{1}{6} = \frac{1}{30}(5)^3 - 10\ln 5 + 10 + B$$

$$\Rightarrow \qquad\qquad B = 10\ln 5 - 10$$

Hence: $\qquad x = \dfrac{t^3}{30} - 10\ln t + 2t + 10\ln 5 - 10, \quad t \geqslant 5$

When $t = 10$:

$$x = \frac{(10)^3}{30} - 10\ln 10 + 20 + 10\ln 5 - 10$$

$$= 43\frac{1}{3} - 10\ln 2$$

The displacement of P from O when $t = 10$ is $(43\frac{1}{3} - 10\ln 2)$ metres.

Example 5

A particle P moves in a straight line with acceleration $(\sin \omega t)$ m s^{-2} at time t seconds where ω is constant. Initially the particle is at rest and its displacement from a fixed point O on the line is 1 metre.

(a) Find the velocity of P at any time t.
(b) Find the displacement of P from O at any time t.
(c) Find the value of t when P next comes to instantaneous rest.

(a) Since: $\qquad\qquad a = \dfrac{\mathrm{d}v}{\mathrm{d}t} = \sin \omega t$

so: $\qquad\qquad v = \displaystyle\int \sin \omega t \, \mathrm{d}t + c$

$$= -\frac{1}{\omega}\cos \omega t + c$$

where c is an arbitrary constant.

Since $v = 0$ when $t = 0$:

$$0 = -\frac{1}{\omega}\cos 0 + c$$

So:
$$c = \frac{1}{\omega}, \text{ as } \cos 0 = 1,$$

and:
$$v = \frac{1}{\omega}(1 - \cos \omega t)$$

(b) As:
$$v = \frac{dx}{dt} = \frac{1}{\omega}(1 - \cos \omega t)$$

so:
$$x = \frac{1}{\omega}\int(1 - \cos \omega t)dt + k$$

$$= \frac{1}{\omega}(t - \frac{1}{\omega}\sin \omega t) + k$$

where k is an arbitrary constant.

Since $x = 1$ when $t = 0$:

$$1 = 0 - 0 + k$$

So:
$$k = 1$$

and:
$$x = \frac{t}{\omega} - \frac{1}{\omega^2}\sin \omega t + 1$$

(c) P comes to rest when $v = 0$, that is:

$$\frac{1}{\omega}(1 - \cos \omega t) = 0$$

$\Rightarrow$
$$\cos \omega t = 1$$

$$\omega t = 2m\pi, \text{ where } m \text{ is an integer}$$

So:
$$t = 0, \frac{2\pi}{\omega}, \frac{4\pi}{\omega}, \ldots$$

P next comes to rest when $t = \frac{2\pi}{\omega}$ s.

Example 6

A particle P moves in a straight line with acceleration $(\cos \pi t)\,\mathrm{m\,s}^{-2}$ at time t seconds. The particle starts from rest at $t = 0$. Obtain the distance travelled by P in the time interval from $t = 1$ to $t = 2$.

As:
$$a = \frac{dv}{dt} = \cos \pi t$$

so:
$$v = \int \cos \pi t \, dt + c$$

$$= \frac{1}{\pi}\sin \pi t + c$$

Since $v = 0$ when $t = 0$:

$$0 = \frac{1}{\pi} \sin 0 + c$$

So: $\qquad\qquad\qquad c = 0$, as $\sin 0 = 0$,

and: $\qquad\qquad\qquad v = \frac{1}{\pi} \sin \pi t$

Before proceeding it is useful to sketch v against t. The graph of $y = \sin x$ is shown in Book P1, chapter 7 and using this gives:

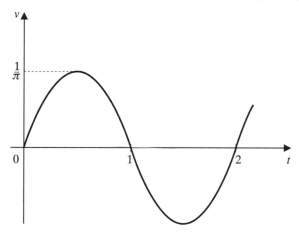

For $1 \leqslant t \leqslant 2$ the velocity is never positive which means that P is moving in the same direction throughout this interval – in the negative x direction.

To obtain information about displacements and distances integrate again with respect to t.

As: $\qquad\qquad\qquad v = \dfrac{\mathrm{d}x}{\mathrm{d}t} = \dfrac{1}{\pi} \sin \pi t$

so: $\qquad\qquad\qquad x = \dfrac{1}{\pi} \displaystyle\int (\sin \pi t)\mathrm{d}t + k$

$$= -\frac{1}{\pi^2} \cos \pi t + k$$

where k is an arbitrary constant.

As no information is given about the displacement at some fixed time the value of k cannot be found. However, as you will see below, it is not necessary to do so.

When $t = 1$ the displacement:

$$x_1 = -\frac{1}{\pi^2} \cos \pi + k$$

$$= +\frac{1}{\pi^2} + k, \text{ as } \cos \pi = -1$$

When $t = 2$ the displacement:

$$x_2 = -\frac{1}{\pi^2} \cos 2\pi + k$$

$$= -\frac{1}{\pi^2} + k, \text{ as } \cos 2\pi = 1$$

As the motion in the given interval is in a constant direction, the distance travelled by P is given by

$$|x_2 - x_1| = \left| \left(-\frac{1}{\pi^2} + k \right) - \left(\frac{1}{\pi^2} + k \right) \right|$$

$$= \frac{2}{\pi^2}$$

Notice that k does not appear in the answer as it cancels out.

The distance travelled by P in the time interval from $t = 1$ to $t = 2$ is $\frac{2}{\pi^2}$ metres.

Example 7

A particle P moves in a straight line with acceleration $(12t) \, \text{m s}^{-2}$ at time t seconds. Its displacement at this time is x metres. Given that $x = 5$ when $t = 1$ and $x = 59$ when $t = 3$, find an expression for x as a function of t and also the value of x when $t = 4$.

As:
$$a = \frac{dv}{dt} = 12t$$

so:
$$v = \int 12t \, dt + A$$

$$= 12 \left(\frac{t^2}{2} \right) + A = 6t^2 + A$$

where A is an arbitrary constant.

No information is given about the value of v at any time so you cannot find A yet.

Integrating again with respect to t:

$$x = \int (6t^2 + A)\mathrm{d}t + B$$

$$= 6\left(\frac{t^3}{3}\right) + At + B$$

$$= 2t^3 + At + B$$

where B is another arbitrary constant.

Since $x = 5$ when $t = 1$:

$$5 = 2 + A + B$$

So: $\qquad\qquad A + B = 3 \qquad\qquad (1)$

Since $x = 59$ when $t = 3$:

$$59 = 2(3)^3 + 3A + B$$
$$= 54 + 3A + B$$

So: $\qquad\qquad 3A + B = 5 \qquad\qquad (2)$

Solve equations (1) and (2) to obtain A and B.

Subtracting equation (1) from equation (2) gives:

$$2A = 2$$

So: $\qquad\qquad A = 1$

Substituting into equation (1):

$$1 + B = 3$$
$$\Rightarrow B = 2$$

Hence: $\qquad\qquad x = 2t^3 + t + 2$

When $t = 4$: $\qquad x = 2(4)^3 + 4 + 2$
$$= 128 + 4 + 2$$
$$x = 134$$

When $t = 4$ s, $x = 134$ m.

Exercise 1A

1 A particle P moves in a straight line with an acceleration $4t\,\mathrm{m\,s}^{-1}$ at time t seconds. When $t = 0$ the velocity of P is $3\,\mathrm{m\,s}^{-1}$. Find the velocity of P when $t = 3$.

2 A particle P moves in a straight line with an acceleration $6t\,\mathrm{m\,s}^{-2}$ at time t seconds. The particle starts from rest. Find the velocity of P when $t = 4$.

3 A particle P moves in a straight line with an acceleration $2t\,\mathrm{m\,s}^{-2}$ at time t seconds. It starts from rest at O. Find the velocity of P and the distance of P from O when $t = 3$.

4 A particle P moves in a straight line with acceleration $\left(\dfrac{3t^2}{2}\right)\,\mathrm{m\,s}^{-2}$ at time t seconds. The particle passes through the point O when $t = 2$ with velocity $1\,\mathrm{m\,s}^{-1}$. Find the velocity of P when $t = 4$ and the distance of P from O at this time.

5 A particle P starts from rest at O at time $t = 0$ and moves with variable acceleration $a\,\mathrm{m\,s}^{-2}$ where

$$a = 2t, \quad 0 \leqslant t < 3$$
$$a = -2, \qquad t \geqslant 3$$

where t is measured in seconds. Find the value of t when the particle comes to rest and the displacement of P from O at this time.

6 A particle P moves in a straight line with an acceleration $(3t - 1)\,\mathrm{m\,s}^{-2}$ at time t seconds. The particle has a velocity of $3\,\mathrm{m\,s}^{-1}$ when $t = 2$. Find the velocity of P when $t = 4$.

7 A particle P moves in a straight line with an acceleration $(3t - 6)\,\mathrm{m\,s}^{-2}$ at time t seconds. Initially P is at a distance of $2\,\mathrm{m}$ from O and moving with a velocity of $4\frac{1}{2}\,\mathrm{m\,s}^{-1}$.
   (a) Find the values of t when the velocity is zero.
   (b) Find the displacement of the particle from O when $t = 4$.

8 A particle P moves in a straight line with acceleration $(\sin 2t)\,\mathrm{m\,s}^{-2}$ at time t seconds. The particle is initially at rest at O. Find the velocity of P and its displacement from O when $t = \dfrac{\pi}{2}$.

9 A particle moves in a straight line and its acceleration after t seconds is $6t\,\mathrm{m\,s}^{-2}$. Its displacement at time t is x metres. Given that $x = 5$ when $t = 1$ and $x = 14$ when $t = 2$, find an expression for x as a function of t and the value of x when $t = 3$.

10 A particle P moves along Ox with an acceleration $(\mathrm{e}^{-t})\,\mathrm{m\,s}^{-2}$ at time t seconds. Given that at $t = 0$ the speed of P is $4\,\mathrm{m\,s}^{-1}$ in

the direction Ox, and its displacement from O is $5\,\mathrm{m}$, find the displacement of P from O as a function of t.

11 A particle P moves along Ox with an acceleration

$\dfrac{-1}{(4+t)^{1/2}}\,\mathrm{m\,s}^{-2}$ at time t seconds. Given that at $t=0$ the speed of P is $2\,\mathrm{m\,s}^{-1}$ in the direction Ox, find the value of t when the speed of P is zero.

12 A particle P moves along Ox with an acceleration $\dfrac{3}{(1+t)^2}\,\mathrm{m\,s}^{-2}$ at time t seconds. The particle starts from rest at O at time $t=0$. Find the displacement of P from O when $t=6$.

13 A particle P moves in a straight line with acceleration, at time t seconds, proportional to $(t+t_0)^{-3}$, where t_0 is positive and constant. The initial speed of P at $t=0$ is $u\,\mathrm{m\,s}^{-1}$.
Show that the speed of P approaches a limiting value as $t\to\infty$.
Given that this limiting value is $2u\,\mathrm{m\,s}^{-1}$, show that at time t seconds the particle will have travelled a distance

$$\frac{ut(2t+t_0)}{(t+t_0)}\ \text{metres.}$$

1.2 Acceleration depending on displacement (x)

Consider a particle P moving in a straight line. Suppose that at time t seconds the displacement of P from a fixed point O in the line is x metres, its velocity is $v\,\mathrm{m\,s}^{-1}$ and its acceleration is $a\,\mathrm{m\,s}^{-2}$.

In section 1.1 the equation $a=\dfrac{\mathrm{d}v}{\mathrm{d}t}$ was used. When the acceleration is a function of the displacement:

$$a=\mathrm{F}(x)$$

so:
$$\frac{\mathrm{d}v}{\mathrm{d}t}=\mathrm{F}(x)$$

This equation contains 3 variables: x, t and v. In order to be able to solve this equation an alternative form for the acceleration must be obtained.

Write: $a = \dfrac{\mathrm{d}v}{\mathrm{d}t} = \dfrac{\mathrm{d}v}{\mathrm{d}x}\dfrac{\mathrm{d}x}{\mathrm{d}t}$ by the chain rule (Book P2, chapter 8)

Since $\dfrac{\mathrm{d}x}{\mathrm{d}t} = v$ this gives:

∎
$$a = v\,\frac{\mathrm{d}v}{\mathrm{d}x}$$

Since $\dfrac{\mathrm{d}}{\mathrm{d}x}(v^2) = 2v\dfrac{\mathrm{d}v}{\mathrm{d}x}$ the acceleration may also be written:

∎
$$a = \frac{\mathrm{d}}{\mathrm{d}x}\left(\tfrac{1}{2}v^2\right)$$

Example 8

A particle P moves along the positive x-axis and its acceleration is $3x\,\mathrm{m\,s}^{-2}$, when its displacement from the origin O is x metres. Given that when $x = 0$ the velocity of P is $8\,\mathrm{m\,s}^{-1}$ in the direction Ox, obtain an expression for v in terms of x.

As:
$$a = \frac{\mathrm{d}}{\mathrm{d}x}\left(\tfrac{1}{2}v^2\right)$$

so:
$$\frac{\mathrm{d}}{\mathrm{d}x}\left(\tfrac{1}{2}v^2\right) = 3x$$

Integrating with respect to x gives:

$$\tfrac{1}{2}v^2 = \int 3x\,\mathrm{d}x + c$$
$$= 3 \times \frac{x^2}{2} + c = 1\tfrac{1}{2}x^2 + c$$

where c is an arbitrary constant.

Since $v = 8$ when $x = 0$:

$$\tfrac{1}{2}(8)^2 = 0 + c$$

So: $\qquad\qquad c = 32$

and: $\qquad\qquad \tfrac{1}{2}v^2 = 1\tfrac{1}{2}x^2 + 32$

or: $\qquad\qquad v^2 = 3x^2 + 64$

Taking square roots:

$$v = \pm\sqrt{(3x^2 + 64)}$$

The acceleration given in the question is positive for $x > 0$ so that P will always be moving in the direction Ox and v will always be positive and increasing.

So: $\qquad\qquad v = \sqrt{(3x^2 + 64)}$

Example 9

A particle P moves along a straight line and its acceleration when it has a displacement of x metres from O, a fixed point in the line, is of magnitude $8x\,\mathrm{m\,s}^{-2}$ and directed towards O. The velocity of the particle is $v\,\mathrm{m\,s}^{-1}$ and initially $v = 0$ when $x = 2$.

(a) Obtain an expression for v^2 as a function of x.
(b) Find the values of x for which $v = 0$.
(c) Find the maximum speed of P.

(a) You are given $a = \dfrac{\mathrm{d}}{\mathrm{d}x}\left(\dfrac{1}{2}v^2\right) = -8x$

The minus sign shows that the acceleration is directed *towards* O.

So integrating with respect to x gives:

$$\frac{1}{2}v^2 = -\int 8x\,\mathrm{d}x + c$$

$$= -8 \times \frac{x^2}{2} + c = -4x^2 + c$$

where c is an arbitrary constant.

Since $v = 0$ when $x = 2$:

$$0 = -4(2)^2 + c$$

So: $\qquad\qquad c = 16$

and: $\qquad\qquad \frac{1}{2}v^2 = -4x^2 + 16$

or: $\qquad\qquad v^2 = -8x^2 + 32 = 32 - 8x^2 \qquad\qquad (1)$

(b) Substituting $v = 0$ in equation (1) gives:

$$32 - 8x^2 = 0$$

or: $\qquad\qquad 4 - x^2 = 0$

so: $\qquad\qquad x^2 = 4$

$\Rightarrow \qquad\qquad x = \pm 2$

(c) The stationary values of v occur when:

$$a = \frac{\mathrm{d}v}{\mathrm{d}t} = v\frac{\mathrm{d}v}{\mathrm{d}x} = \frac{\mathrm{d}}{\mathrm{d}x}\left(\frac{1}{2}v^2\right) = 0$$

That is when: $\qquad\qquad 8x = 0$

$\Rightarrow \qquad\qquad x = 0$

When $x = 0$: $\qquad\qquad v^2 = 32$

$\Rightarrow \qquad\qquad v = \pm\sqrt{32} = \pm 4\sqrt{2}$

The maximum value of $|v|$ is $4\sqrt{2}$ and so the maximum speed of P is $4\sqrt{2}\,\mathrm{m\,s}^{-1}$.

Example 10

A particle P moves along the line Ox. When its displacement from O is x metres, its acceleration is of magnitude $\left(\dfrac{k^2}{2x^2}\right) \mathrm{m\,s^{-2}}$ and directed towards O. Given that when $x = \frac{1}{4}a$ the velocity of P is $\left(\dfrac{3k^2}{a}\right)^{\frac{1}{2}} \mathrm{m\,s^{-1}}$ in the direction Ox, find the value of x when P comes to rest.

In this case:

$$a = \frac{\mathrm{d}}{\mathrm{d}x}\left(\frac{1}{2}v^2\right) = -\frac{k^2}{2x^2}$$

as the acceleration is towards O.

So integrating with respect to x:

$$\begin{aligned}
\frac{1}{2}v^2 &= -\int \frac{k^2}{2x^2}\,\mathrm{d}x + c \\
&= -\frac{k^2}{2}\left(\frac{x^{-1}}{-1}\right) + c \\
&= \frac{k^2}{2x} + c
\end{aligned}$$

where c is an arbitrary constant.

As $v = \left(\dfrac{3k^2}{a}\right)^{\frac{1}{2}}$ when $x = \frac{1}{4}a$,

so:

$$\frac{1}{2}\left(\frac{3k^2}{a}\right) = \frac{k^2}{2\left(\frac{a}{4}\right)} + c$$

$\Rightarrow$

$$c = \frac{3k^2}{2a} - \frac{4k^2}{2a} = -\frac{k^2}{2a}$$

and:

$$\frac{1}{2}v^2 = \frac{k^2}{2x} - \frac{k^2}{2a}$$

or:

$$v^2 = k^2\left(\frac{1}{x} - \frac{1}{a}\right)$$

P comes to rest when $v = 0$, that is when:

$$\frac{1}{x} - \frac{1}{a} = 0$$

$\Rightarrow$

$$x = a$$

When P comes to rest, $x = a$.

Example 11

A particle P moves along the line Ox. When its displacement from O is x metres, its acceleration is of magnitude $\left(\dfrac{8}{x^2}\right)$ m s^{-2} and directed towards O. When $x = 1$ the velocity of P is 5 m s^{-1} in the direction Ox. Show that the velocity of P tends to 3 as x tends to infinity.

In this case:
$$a = \frac{\mathrm{d}}{\mathrm{d}x}\left(\frac{1}{2}v^2\right) = -\frac{8}{x^2}$$

Integrating with respect to x:
$$\frac{1}{2}v^2 = \int -\frac{8}{x^2}\ \mathrm{d}x + c = -8\left(\frac{x^{-1}}{-1}\right) + c$$
$$= \frac{8}{x} + c$$

where c is an arbitrary constant.

Since $v = 5$ when $x = 1$:
$$\frac{1}{2}(5)^2 = \frac{8}{1} + c$$

So:
$$c = 4\tfrac{1}{2}$$

and:
$$\frac{1}{2}v^2 = \frac{8}{x} + 4\tfrac{1}{2}$$

or:
$$v^2 = \frac{16}{x} + 9$$

As x tends to infinity, $\dfrac{16}{x}$ tends to zero and so $v^2 \to 9$.

Hence $v \to 3$, since v is initially positive.

Exercise 1B

1 A particle P moves along the positive x-axis with constant acceleration f m s^{-2} $(f > 0)$. When P is at the origin its velocity is u m s^{-1} in the direction Ox. Find the velocity v m s^{-1} in terms of the displacement x metres from O.

2 A particle P moves along a straight line. Its acceleration when it has a displacement of x metres from O, a fixed point in the line, is of magnitude $4x$ m s^{-2} and directed towards O. Given that the velocity of P is zero when $x = 5$, find the speed of P when $x = 3$.

3 A particle P moves along a straight line and when its distance from a fixed point O is x metres, its retardation is $(20 + 2x)\,\mathrm{m\,s^{-2}}$. Given that the speed of P is $40\,\mathrm{m\,s^{-1}}$ when $x = 0$, obtain an expression for v in terms of x. Find the value of x when P comes to rest.

4 The acceleration of a particle P, moving on a straight line, is $\lambda x^2\,\mathrm{m\,s^{-2}}$ $(\lambda > 0)$ when P has a displacement x metres from O, a fixed point in the line. The acceleration is towards O. The speed of P is zero when $x = a$. Find the speed of P when $x = 0$.

5 When a particle P has a displacement of x metres from a fixed point O in a straight line, its acceleration is $\dfrac{k}{x}\,\mathrm{m\,s^{-2}}$ $(k > 0)$ away from O. It has zero velocity when $x = 1$. Find x in terms of v and k.

6 The acceleration of a particle P, moving in a straight line, is $4x^3\,\mathrm{m\,s^{-2}}$ when P has a displacement x metres from O, a fixed point in the line. The acceleration is away from O. Given that $v = 2$ when $x = 0$, obtain the value of x when $v = 6$.

7 A particle P moves along the positive x-axis. Its acceleration is $(x + 3)\,\mathrm{m\,s^{-2}}$ when its displacement from the origin O is x metres. Given that initially, when $t = 0$, the velocity of P is $3\,\mathrm{m\,s^{-1}}$ in the direction Ox and $x = 0$, obtain (a) the speed $v\,\mathrm{m\,s^{-1}}$ of P as a function of x (b) x as a function of t.

8 A particle moves along the positive x-axis. When it has a displacement of x metres from O its acceleration is of magnitude $\dfrac{4}{x^3}\,\mathrm{m\,s^{-2}}$ and directed towards O. Given that initially, when $t = 0$, the velocity of the particle is $2\,\mathrm{m\,s^{-1}}$ in the direction Ox and $x = 1$, obtain (a) v as a function of x (b) x as a function of t.

9 A particle P moves along the positive x-axis and its acceleration is $(e^{2x})\,\mathrm{m\,s^{-2}}$ when its displacement from O is x metres. Given that initially, when $t = 0$, the velocity of P is $u\,\mathrm{m\,s^{-1}}$ towards O and $x = \ln u$, obtain (a) v as a function of x (b) x as a function of t.

10 A particle moves on the positive x-axis. When its displacement from O is x metres, its acceleration is of magnitude $\dfrac{1}{x^3}\,\mathrm{m\,s^{-2}}$ and directed towards O. Given that initially, when $t = 0$, the particle is at rest with $x = 1$, find the time the particle takes to reach $x = \frac{1}{4}$.

11 A particle P moves on the positive x-axis and its acceleration, when it has a displacement of x metres from O, is $(4 \sin x)\,\mathrm{m\,s}^{-2}$ in the direction Ox. Given that when $x = 0$ the velocity of P is $6\,\mathrm{m\,s}^{-1}$ in the direction Ox, find the maximum speed of P.

12 A particle P moves on the positive x-axis and its acceleration, when it has a displacement of x metres from O, is $(\mathrm{e}^{2x})\,\mathrm{m\,s}^{-2}$. Given that initially, when $t = 0$, the speed of P is $1\,\mathrm{m\,s}^{-1}$ in the direction Ox and $x = 0$, obtain x as a function of t and show that $x \to \infty$ as $t \to 1$.

13 A particle P moves on the positive x-axis. When it has a displacement of x metres from O, it has acceleration $3\sqrt{x}\,\mathrm{m\,s}^{-2}$ in the direction Ox. When $t = 0$ the particle starts from rest at O. Show that $x = \frac{1}{16}t^4$.

14 A particle P moves on the positive x-axis and its acceleration, when it has a displacement of x metres from O, is $2\sin^2 x\,\mathrm{m\,s}^{-2}$. When $x = \dfrac{\pi}{4}$ the velocity of P is $1\,\mathrm{m\,s}^{-1}$ in the direction Ox. Find the velocity of P when $x = \dfrac{\pi}{3}$.

SUMMARY OF KEY POINTS

1 For a particle P travelling in a straight line, which at time t seconds has a displacement x metres from a fixed point O of the line, the acceleration $a\,\mathrm{m\,s}^{-2}$ and the velocity $v\,\mathrm{m\,s}^{-1}$ as functions of time are given by:

$$a = \frac{\mathrm{d}v}{\mathrm{d}t}$$

$$v = \frac{\mathrm{d}x}{\mathrm{d}t}$$

$$a = \frac{\mathrm{d}^2x}{\mathrm{d}t^2}$$

2 When the acceleration is a function of the displacement then:

$$a = v\frac{\mathrm{d}v}{\mathrm{d}x} = \frac{\mathrm{d}}{\mathrm{d}x}\left(\frac{1}{2}v^2\right)$$

Further dynamics

2

Dynamics is the study of the relationship between the resultant force acting on a particle and the motion of that particle. The study of dynamics was begun in Book M1 but was restricted to constant forces. This work will now be extended to include some cases of variable forces.

2.1 Newton's laws for a particle moving in a straight line when the applied force is variable

Chapter 1 looked at the kinematics of a particle moving in a straight line when the acceleration was changing during the motion. Consideration of Newton's second law, $\mathbf{F} = m\mathbf{a}$, will show you that for a particle of constant mass, the acceleration can only be variable if the resultant force acting on the particle is variable.

Consider a particle of mass m moving under the influence of a force F which is a function of time.

That is: $$F = \mathrm{F}(t)$$

Using Newton's second law, $F = ma$, gives:

$$ma = \mathrm{F}(t)$$

As: $$a = \frac{\mathrm{d}v}{\mathrm{d}t}$$

It follows that: $$m\frac{\mathrm{d}v}{\mathrm{d}t} = \mathrm{F}(t)$$

and so, integrating with respect to time gives:

$$mv = \int \mathrm{F}(t)\mathrm{d}t + c$$

where c is an arbitrary constant.

Similarly, if F is a function of the displacement, that is:

$$F = G(x)$$

then using Newton's second law gives:

$$ma = G(x)$$

In chapter 1 you were shown that

$$a = \frac{d}{dx}\left(\frac{1}{2}v^2\right)$$

and so:
$$m\frac{d}{dx}\left(\frac{1}{2}v^2\right) = G(x)$$

Integrating with respect to x gives:

$$\frac{1}{2}mv^2 = \int G(x)dx + k$$

where k is an arbitrary constant.

Example 1

A particle P of mass $2\,kg$ is moving along a straight line. P is initially at rest at a point O on the line.
The force acting on P has magnitude $(3t^2 + e^t)\,N$ at time $t\,s$ and acts in the direction OP. Calculate:

(a) the speed of P when $t = 2$
(b) the distance OP when $t = 4$.

(a) Using Newton's second law, $F = ma$, with $F = (3t^2 + e^t)\,N$ and $m = 2\,kg$ gives:

$$3t^2 + e^t = 2a$$

Using:
$$a = \frac{dv}{dt}$$

gives:
$$2\frac{dv}{dt} = 3t^2 + e^t$$

Integrating with respect to t gives:

$$2v = \frac{3}{3}t^3 + e^t + c$$
$$= t^3 + e^t + c$$

As $v = 0$ when $t = 0$:

$$0 = 0 + 1 + c$$

So:
$$c = -1$$

and:
$$2v = t^3 + e^t - 1 \qquad (1)$$

Substituting $t = 2$ gives:
$$2v = 8 + e^2 - 1$$
$$v = \tfrac{7}{2} + \tfrac{1}{2} e^2$$

The speed of P when $t = 2$ is $(3\tfrac{1}{2} + \tfrac{1}{2} e^2)\,\mathrm{m\,s^{-1}}$.

(b) From equation (1) above, $2v = t^3 + e^t - 1$.

Using $v = \dfrac{\mathrm{d}x}{\mathrm{d}t}$ gives:
$$2 \frac{\mathrm{d}x}{\mathrm{d}t} = t^3 + e^t - 1$$

Integrating with respect to t gives:

$$2x = \frac{1}{4} t^4 + e^t - t + K$$

As $x = 0$ when $t = 0$:

$$0 = e^0 + K$$

So:
$$K = -1$$

And:
$$2x = \frac{1}{4} t^4 + e^t - t - 1$$

Substituting $t = 4$ gives:
$$2x = \frac{1}{4} \times 4^4 + e^4 - 4 - 1$$

$$2x = 59 + e^4$$

$$x = \frac{59}{2} + \frac{1}{2} e^4$$

The distance OP when $t = 4$ is $(29\tfrac{1}{2} + \tfrac{1}{2} e^4)\,\mathrm{m}$.

Example 2

A particle P of mass $0.5\,\mathrm{kg}$ is moving in a straight line. When P is at a distance $x\,\mathrm{m}$ from a fixed point O on the line, the force acting on it is of magnitude $(6 - \cos 3x)\,\mathrm{N}$ and acts in the direction OP. Given that P passes through O with speed $5\,\mathrm{m\,s^{-1}}$, calculate the speed of P when $x = 2$.

Using Newton's second law, $F = ma$, with $F = (6 - \cos 3x)\,\mathrm{N}$ and $m = 0.5\,\mathrm{kg}$ gives:

$$6 - \cos 3x = 0.5\,a$$

As the force is a function of x, use:

$$a = \frac{\mathrm{d}}{\mathrm{d}x}\left(\frac{1}{2} v^2\right)$$

and obtain: $0.5 \dfrac{\mathrm{d}}{\mathrm{d}x}\left(\dfrac{1}{2} v^2\right) = 6 - \cos 3x$

Integrating with respect to x gives:

$$0.5 \times \frac{1}{2} v^2 = 6x - \frac{1}{3} \sin 3x + c$$

Since $v = 5$ when $x = 0$:

$$\frac{1}{2} \times \frac{1}{2} \times 5^2 = 0 - 0 + c$$

So:
$$c = \frac{25}{4}$$

And so:
$$\frac{1}{4} v^2 = 6x - \frac{1}{3} \sin 3x + \frac{25}{4}$$

Substituting $x = 2$ gives:

$$\frac{1}{4} v^2 = 6 \times 2 - \frac{1}{3} \sin (3 \times 2) + \frac{25}{4}$$
$$v = 8.565$$

The speed of P when $x = 2$ is $8.57 \, \text{m s}^{-1}$.

Impulse and momentum

The impulse–momentum equation $\mathbf{F}t = m\mathbf{v} - m\mathbf{u}$ was derived in Book M1 chapter 5 for a constant force $\mathbf{F}$ acting on a particle of constant mass m. For a variable force, $F(t)$, Newton's second law, $F(t) = ma$, can be written:

$$F(t) = m \frac{dv}{dt}$$

Suppose the particle has constant mass and a velocity U at time t_1 and a velocity V at time t_2. Then integrating with respect to t over the interval from t_1 to t_2 gives:

$$\int_{t_1}^{t_2} F(t)dt = m \int_U^V 1 \, dv$$
$$= m[v]_U^V$$
$$= m(V - U)$$
$$= mV - mU$$

You should recognise $mV - mU$ as the change of momentum of the particle. The quantity $\int_{t_1}^{t_2} F(t) \, dt$ is the impulse of the variable force on the particle. Thus the equation:

impulse = change of momentum

is still valid but for a variable force the impulse is given by:

$$\text{impulse} = \int_{t_1}^{t_2} F\,dt$$

Example 3

A stone of mass 0.25 kg is free to move on a smooth horizontal floor. At time t seconds, $t \geqslant 0$, a horizontal force of magnitude $3t^2$ N and constant direction acts on the stone. When $t = 0$ the stone has speed $8\,\text{m s}^{-1}$ in the same direction as the force. When $t = T$ the speed of the stone is $40\,\text{m s}^{-1}$.

   (a) Choose suitable models for the stone and the floor so that you can find the value of T.
   (b) Calculate the value of T.

(a) The stone can be assumed to be small and hence modelled as a particle.

The floor is said to be smooth and horizontal and so can be modelled as a smooth horizontal surface.

(b) Using $\displaystyle\int_{t_1}^{t_2} F\,dt = mV - mU$

with $F = 3t^2$ N, $V = 40$ m s^{-1}, $U = 8$ m s^{-1}, $t_1 = 0$ s, $t_2 = T$ s and $m = 0.25$ kg gives:

$$\int_{0}^{T} 3t^2\,dt = 0.25 \times 40 - 0.25 \times 8$$
$$[t^3]_{0}^{T} = 10 - 2$$
$$T^3 = 8$$
$$T = 2$$

Work and energy

When a *constant* force **F** acts on a particle of mass m and moves it through a distance s in the direction of the force, the work done by the force is defined by:

$$\textit{Work done} = F \times s$$

In chapter 5 of Book M1 you saw that:

$$Fs = \tfrac{1}{2}mv^2 - \tfrac{1}{2}mu^2$$

where u and v are the initial and final speeds respectively of the particle.

When the force F acting on the particle is *variable* and a function of x, using Newton's second law with $a = v \dfrac{dv}{dx}$ (see chapter 1) gives:

$$F(x) = mv \frac{dv}{dx} \qquad (1)$$

If the action of the force causes the particle to increase its speed from U to V while moving from the point where $x = x_1$, to the point where $x = x_2$, then integrating equation (1) with respect to x gives:

$$\int_{x_1}^{x_2} F(x)dx = m \int_U^V v\,dv = m \left[\frac{1}{2} v^2 \right]_U^V$$

$$\int_{x_1}^{x_2} F(x)dx = \frac{1}{2} mV^2 - \frac{1}{2} mU^2 \qquad (2)$$

The right hand side of equation (2) is the increase in the kinetic energy of the particle and the left hand side is the work done by the variable force. Thus the equation:

work done by force = increase in kinetic energy

is still valid but for a variable force $F(x)$ the work done by the force is given by:

■ $$\textbf{work done} = \int_{x_1}^{x_2} F(x)dx$$

Example 4

A particle P of mass $0.5\,kg$ is moving along the x-axis. When P is a distance $x\,m$ from the origin, O, a force of magnitude $(10 + 5e^{-x})\,N$ acts on P in the direction of the positive x-axis. Calculate the work done by the force in moving P from O to the point A where $x = 2$. Given that P passes through the origin with speed $3\,m\,s^{-1}$ calculate the speed of P as it passes through A.

Using: $$\text{work done} = \int_{x_1}^{x_2} F\,dx$$

with $x_1 = 0$, $x_2 = 2$ and $F = (10 + 5e^{-x})$ gives:

$$
\begin{aligned}
\text{work done} &= \int_0^2 (10 + 5e^{-x})dx \\
&= [10x - 5e^{-x}]_0^2 \\
&= (20 - 5e^{-2}) - (0 - 5e^0) \\
&= 20 - 5e^{-2} + 5 \\
&= 25 - 5e^{-2} \\
&= 24.32
\end{aligned}
$$

The work done is $24.3\,J$.

Let the speed of P as it passes through A be $v\,\mathrm{m\,s}^{-1}$.

Using: work done $=$ increase in kinetic energy

$$= \tfrac{1}{2}\,mv^2 - \tfrac{1}{2}\,mu^2$$

with work done $= 24.32\,\mathrm{J}$, $m = 0.5\,\mathrm{kg}$, $u = 3\,\mathrm{m\,s}^{-1}$ gives:

$$24.32 = \tfrac{1}{2} \times 0.5v^2 - \tfrac{1}{2} \times 0.5 \times 9$$

$$\tfrac{1}{2} \times 0.5v^2 = 24.32 + \tfrac{1}{2} \times 0.5 \times 9$$

$$v^2 = \frac{24.32 + \tfrac{1}{2} \times 0.5 \times 9}{\tfrac{1}{2} \times 0.5}$$

$$v = 10.30$$

The speed of P as it passes through A is $10.3\ \mathrm{m\,s}^{-1}$.

Exercise 2A

1 A particle P of mass $0.5\,\mathrm{kg}$ is moving in a straight line. P is initially at rest at a point O of the line. The force acting on P has magnitude $4\sin t\,\mathrm{N}$ at time $t\,\mathrm{s}$ and acts in the direction OP. Calculate (a) the speed of P when $t = 1$ (b) the speed of P when $t = 3$ (c) the time when P first comes to instantaneous rest (d) the distance OP when $t = 3$ (e) the distance OP when P first comes to rest.

2 A particle P of mass $1.5\,\mathrm{kg}$ is moving in a straight line. At time $t\,\mathrm{s}$ after it passes through a fixed point O in that line the distance OP is $x\,\mathrm{m}$ and the force acting on P has magnitude $(3 + 3\mathrm{e}^{-2x})\,\mathrm{N}$ and acts in the direction OP. Given that P passes through O with speed $2\ \mathrm{m\,s}^{-1}$ calculate the speed of P when (a) $x = 2$ (b) $x = 5$.

3 A particle P of mass $0.5\,\mathrm{kg}$ moves along the x-axis under the action of a force whose magnitude at time $t\,\mathrm{s}$ after P passes through the origin is given by $(4t^3 + 3t^2)\,\mathrm{N}$ and which acts in the direction OP. Given that P passes through the origin with velocity $6\ \mathrm{m\,s}^{-1}$ calculate (a) the velocity of P when $t = 2$ (b) the displacement of P when $t = 1$.

4 A particle P of mass $2\,\mathrm{kg}$ moves along the x-axis under the action of a force whose magnitude at time $t\,\mathrm{s}$ after P passes through the origin is $(4 + 12t)\,\mathrm{N}$ and which acts in the direction

PO. P passes through O with a velocity 20 m s^{-1}. Calculate (a) the speed of P when $t = 2$ (b) the displacement of P when $t = 2$ (c) the time when P has velocity $\frac{25}{3} \text{ m s}^{-1}$ (d) the displacement of P when P's velocity is $\frac{25}{3} \text{ m s}^{-1}$.

5 A particle P of mass 1.5 kg moves in a straight line through a fixed point O. At time t s after passing through O the distance of P from O is x m and the force acting on P has magnitude $(3x + 6)$ N and is directed away from O. Given that P passes through O with speed $2\sqrt{2} \text{ m s}^{-1}$ calculate (a) the speed of P when $x = 5$ (b) the value of t when $x = 20$.

6 Calculate the magnitude of the impulse of a force of magnitude F N acting from time t_1 seconds to time t_2 seconds where

   (a) $F = t^3 + 2t \qquad t_1 = 1 \qquad t_2 = 3$

   (b) $F = \dfrac{1}{2t + 4} \qquad t_1 = 0 \quad t_2 = 4$

   (c) $F = 5 + e^{-2t} \quad t_1 = 0 \quad t_2 = 3$

   (d) $F = 3\cos 5t \quad t_1 = 0 \quad t = \dfrac{\pi}{10}$

7 Calculate the work done by a force of magnitude F N directed along the x-axis which moves a particle from $x = x_1$ m to $x = x_2$ m where

   (a) $F = x^2 + 2x \quad x_1 = 0 \qquad x_2 = 5$
   (b) $F = 2 + e^{-x} \quad x_1 = 2 \qquad x_2 = 6$
   (c) $F = \dfrac{2}{x} + 3 \qquad x_1 = 2 \qquad x_2 = 4$

8 A particle P of mass 2 kg is moving in a straight line. P is initially at rest at a point O of the line. The force acting on P has magnitude $(6t + 4)$ N, where t s is the time for which the force has been acting, and is directed along OP. Calculate (a) the impulse exerted by the force between the times $t = 1$ and $t = 3$ (b) the speed of P when $t = 4$. Given that the final speed of the particle is 66 m s^{-1} calculate (c) the time for which the force acts.

9 A particle P of mass 0.8 kg moves in a straight line through a fixed point O. At time t s after passing through O the distance of P from O is x m and the acceleration of P is $(\frac{1}{3}x^2 + x) \text{ m s}^{-2}$. Write down, in terms of x, an expression for the force acting on

P and calculate the work done by that force in moving P from $x = 0$ to $x = 5$.

2.2 Newton's law of gravitation

In Newton's day most scientists studied astronomy. Newton's study of planetary motion led him to propose the following **law of gravitation**:

■ **The force of attraction between two bodies of masses M_1 and M_2 is directly proportional to the product of their masses and inversely proportional to the square of the distance between them.**

This statement can be expressed mathematically by the equation

$$F = \frac{GM_1M_2}{d^2}$$

where G is a constant known as the **constant of gravitation**.

It is because of this force that bodies fall to the earth and the moon orbits the earth.

The numerical value of G was first determined approximately by Henry Cavendish in 1798. Scientists have since improved Cavendish's approximation and the numerical approximation used today, when the force is measured in newtons, the distance in metres and the masses in kilograms, is 6.67×10^{-11}.

Units of G

From the equation: $F = G \dfrac{M_1M_2}{d^2}$

it follows that: $G = \dfrac{Fd^2}{M_1M_2} = Fd^2(M_1M_2)^{-1}$ (1)

With the usual S.I. units F is in newtons, d is in metres and M_1 and M_2 are in kilograms.

Using Newton's second law, $\mathbf{F} = m\mathbf{a}$, and the fact that the units on both sides of an equation must be the same, it follows that 1 newton must be $1 \, \text{kg} \, \text{m} \, \text{s}^{-2}$.

The units on both sides of equation (1) above must also be the same so the units for G must be:

$$\text{kg} \, \text{m} \, \text{s}^{-2} \, \text{m}^2 \, \text{kg}^{-2}$$

which simplifies to:

$$kg^{-1} m^3 s^{-2}$$

Hence G is $6.67 \times 10^{-11} kg^{-1} m^3 s^{-2}$

Example 5

Two particles of masses $2\,kg$ and $3\,kg$ are $10\,cm$ apart. Determine the magnitude of the gravitational force between them.

Using: $$F = G\,\frac{M_1 M_2}{d^2}$$

with $G = 6.67 \times 10^{-11} kg^{-1} m^3 s^{-2}$, $M_1 = 2\,kg$, $M_2 = 3\,kg$ and $d = 10\,cm = 0.1\,m$ gives:

$$F = \frac{6.67 \times 10^{-11} \times 2 \times 3}{0.1^2}$$
$$F = 4.002 \times 10^{-8}$$

The magnitude of the gravitational force between the particles is $4.0 \times 10^{-8}\,N$.

It can be seen from the above example that the gravitational attraction between everyday objects is very small compared to the other forces acting and so this force may be ignored in calculations.

Relationship between G and g

Consider a particle of mass m which is at rest on the surface of the earth. Newton proved that the gravitational attraction due to the earth is the same as that of a particle whose mass is equal to the earth's mass and is situated at the centre of the earth. So a particle model may be used for the earth.

Suppose the mass of the earth is $M_e\,kg$ and its radius is $R\,m$. By Newton's law of gravitation the magnitude of the force of attraction between the particle and the earth is:

$$F = \frac{Gm M_e}{R^2} \qquad (1)$$

However, as you saw in Book M1, the force with which the earth attracts the particle is the weight of the particle, so:

$$F = mg \qquad (2)$$

Combining equations (1) and (2) gives:

$$mg = \frac{Gm M_e}{R^2}$$

Or: $$g = \frac{GM_e}{R^2}$$

Exercise 2B

Whenever a numerical value of g or G is required take
$g = 9.8 \, \text{m s}^{-2}$ and $G = 6.67 \times 10^{-11} \, \text{kg}^{-1} \, \text{m}^3 \, \text{s}^{-2}$.

1 Two particles have masses 0.2 kg and 0.5 kg. Calculate the magnitude of the gravitational attraction between them when they are (a) 2 cm apart (b) 2 m apart (c) 2 km apart.

2 Two identical particles which are 5 cm apart experience a gravitational attraction of magnitude 7.6×10^{-9} N. Calculate the masses of the particles.

3 The acceleration due to the earth's gravity is taken to be $9.8 \, \text{m s}^{-2}$ at sea level. Obtain an estimate for this acceleration at the top of a mountain which is 7 km above sea level. You may assume the earth to be a sphere of radius 6370 km. State clearly any other assumptions you make.

4 Assuming the earth to be a sphere of radius 6370 km, obtain an estimate for the mass of the earth.

5 The moon can be assumed to be a sphere of radius 1740 km and mass 7.34×10^{22} kg. An astronaut of mass 90 kg is making a moon walk. Calculate the magnitude of the force of attraction between the moon and the astronaut.

6 Assuming the moon to be a sphere of radius 1740 km and mass 7.34×10^{22} kg, obtain an estimate for the acceleration due to the moon's gravity. State clearly the model you have used for the moon.

2.3 Elastic springs and strings

In the work considered so far, all strings have been assumed to be inextensible. In fact no string is totally inextensible although many can be assumed to be so. Those strings which can be stretched are called **elastic**.

An **elastic string** is assumed to regain its original or **natural length** when the force causing it to stretch is removed. The extra length resulting from the force acting on an elastic string is called the **extension**. A stretched string will have a **tension**.

Consider an elastic string AB which has end A fixed. When there is no force applied to end B the length of the string is its natural length l and there is no tension in the string. However, when a force $\mathbf{F}$ is applied to the end B, the string will stretch.

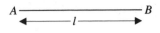

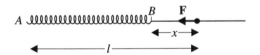

The length of the string is now $l + x$ where x is the extension. The tension $\mathbf{T}$ in the string will be equal in magnitude to the extending force $\mathbf{F}$.

An elastic spring behaves in the same manner as an elastic string when stretched. However, a spring can also be compressed.

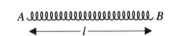

With end A fixed and a force $\mathbf{F}$ applied to end B in the direction BA the spring will be compressed.

The spring will exert a **thrust** $\mathbf{T}$ of magnitude equal to the magnitude of the compressing force $\mathbf{F}$.

If an elastic string is overstretched it may break or become permanently deformed. Similarly if a spring is overstretched or overcompressed it may become deformed and not return to its natural length when released. The point at which permanent deformation takes place is called the **elastic limit**. It will be assumed in what follows that all strings and springs do not reach this limit.

Hooke's law

Consider a simple experiment in which bodies of different masses are attached to the free end of an elastic string, the other end of which is fixed. The resulting extensions of the string may be recorded.

Since the body is in equilibrium $T = mg$ where T is the tension in the string and m is the mass of the body.

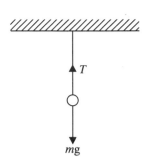

The results of such an experiment show that:

the tension in the string is proportional to the extension produced.

This is known as **Hooke's law**.

Hooke's law is normally used in the form:

■
$$T = \frac{\lambda x}{l}$$

where T is the tension in the string, x is the extension produced and l is the natural length of the string. The constant of proportionality, λ, is called the **modulus of elasticity** or simply the **modulus** of the string, and depends on the nature of the string.

You must remember that all strings are capable of being stretched to some extent although they are often modelled as though this were not the case. Also wires can be stretched and consequently in some situations they are modelled as elastic strings. Most wires would stretch by only a small amount even if the applied force were of large magnitude.

Consideration of the equation:

$$T = \frac{\lambda x}{l}$$

will show you that if a large numerical value for T only produces a small extension x then λ must also be large. The table below gives the modulus of elasticity for wires of diameter 1 mm of some materials.

Material	Aluminium	Copper	Brass	Mild steel
Modulus of elasticity	5.5×10^4	8.8×10^4	7.3×10^4	1.6×10^5

Units must balance in any equation. Both x and l are lengths and consequently are measured in the same units, so λ must have the same units as force. For a force in newtons, λ is also in newtons.

Hooke's law applies equally to springs which are compressed with the thrust being proportional to the compression.

Example 6

An elastic string has natural length $2\,\mathrm{m}$ and modulus $12\,\mathrm{N}$. Calculate (a) the tension in the string when the extension is $0.2\,\mathrm{m}$ (b) the extension of the string when the tension is $8\,\mathrm{N}$.

(a) Using Hooke's law:

$$T = \frac{\lambda x}{l}$$

with $\lambda = 12\,\mathrm{N}$, $l = 2\,\mathrm{m}$, $x = 0.2\,\mathrm{m}$ gives:

$$T = \frac{12 \times 0.2}{2} = 1.2$$

The tension is $1.2\,\mathrm{N}$.

(b) Using Hooke's law:

$$T = \frac{\lambda x}{l}$$

with $\lambda = 12\,\mathrm{N}$, $T = 8\,\mathrm{N}$, $l = 2\,\mathrm{m}$ gives:

$$8 = \frac{12x}{2}$$
$$x = \frac{2 \times 8}{12} = 1.333$$

The extension is $1.33\,\mathrm{m}$.

Example 7

A spring has modulus $30\,\mathrm{N}$ and natural length $0.5\,\mathrm{m}$. Calculate the thrust in the spring when it is compressed a distance of $10\,\mathrm{cm}$.

Let the thrust be $T\,\mathrm{N}$.

Compression $= 10\,\mathrm{cm} = 0.1\,\mathrm{m}$, natural length $= 0.5\,\mathrm{m}$, $\lambda = 30\,\mathrm{N}$

Using Hooke's law:

$$T = \frac{\lambda x}{l}$$

gives:
$$T = 30 \times \frac{0.1}{0.5}$$
$$T = 6$$

The thrust is $6\,\mathrm{N}$.

Equilibrium of a particle suspended on an elastic string or spring

Consider a string or spring with one end fixed to a ceiling and a particle attached to the free end. When the particle is hanging in equilibrium the tension T in the string can be found by resolving vertically, giving:

$$T = mg$$

Using Hooke's law gives:

$$T = \frac{\lambda x}{l}$$

where x is the extension in the string and l is its natural length.

So:

$$mg = \frac{\lambda x}{l}$$

or the extension is given by:

$$x = \frac{mgl}{\lambda}$$

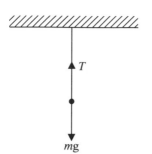

Example 8

A light elastic string of natural length 50 cm and modulus 40 N has one end fixed. A particle of mass 2 kg is attached to the other end and hangs in equilibrium. Calculate the extension in the string.

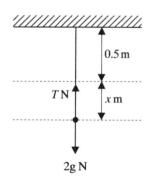

Let the extension be x m.

The particle is in equilibrium so resolving vertically gives:

$$T = 2g$$

Using Hooke's law, $T = \frac{\lambda x}{l}$ with $\lambda = 40$ N, $l = 50$ cm $= 0.5$ m and $T = 2g$ N gives:

$$2g = 40 \times \frac{x}{0.5}$$

$$x = \frac{g}{40} = 0.245$$

The extension in the string is 0.245 m or 24.5 cm.

Example 9

A rope is tied to a branch of a tree. When the rope is hanging freely its length is 2.5 m. When a child of mass 40 kg hangs in equilibrium from the end of the rope, the length of the rope is 2.7 m. Stating clearly any assumptions made, choose suitable models for the rope and the child and hence calculate the modulus of the rope.

The mass of the rope will be small compared with that of the child. The rope can therefore be assumed to be light. It stretches under the weight of the child, so it can be modelled as a light elastic string of natural length 2.5 m.

The child is hanging on to the end of the rope so the contact between the child and the rope is restricted to this point. The child can therefore be modelled as a particle.

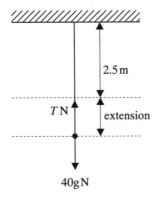

Let the tension in the rope be T N.

As the child is in equilibrium, resolving vertically gives:

$$T = 40\,g \qquad\qquad (1)$$

The extension in the rope is $(2.7 - 2.5) = 0.2$ m.

Using Hooke's law, $T = \dfrac{\lambda x}{l}$, for the rope with $l = 2.5$ m and $x = 0.2$ m gives:

$$T = \lambda \times \frac{0.2}{2.5} \qquad\qquad (2)$$

Eliminating T between equations (1) and (2) gives:

$$\lambda \times \frac{0.2}{2.5} = 40 \times 9.8$$

$$\lambda = 40 \times 9.8 \times \frac{2.5}{0.2}$$

$$\lambda = 4900$$

The modulus of the rope is 4900 N.

Motion of a particle attached to the end of an elastic string or spring

Consider now the motion of a particle attached to one end of an elastic string which is hanging vertically and has its other end fixed. The forces acting on the particle are gravity and the tension in the string. If the particle is at some point other than its equilibrium position there will be a resultant force acting on it and this will produce an acceleration.

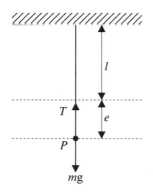

When the particle P is in equilibrium, using Hooke's law gives:

$$T = \frac{\lambda e}{l}$$

where e is the extension.

Hence:

$$e = \frac{Tl}{\lambda}$$

or, as $T = mg$:

$$e = \frac{mgl}{\lambda}$$

When the particle P is at some point where the extension, x, is greater than $\frac{mgl}{\lambda}$ then the tension in the string will be greater than the weight of the particle and the particle will move with an acceleration a.

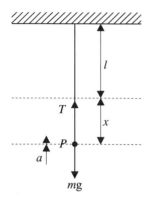

Using Hooke's law in this position gives $T = \dfrac{\lambda x}{l}$.

Using the equation of motion, $F = ma$, gives:

$$T - mg = ma$$

So:

$$\frac{\lambda x}{l} - mg = ma$$

or:

$$a = \frac{\lambda x}{ml} - g$$

The resulting acceleration depends on the extension in the string and is constantly changing.

Example 10

A particle of mass 1 kg is attached to one end A of a light elastic string AB of modulus 25 N and natural length 0.5 m. The other end B of the string is fixed and A is held at a distance of 0.75 m vertically below B and then released. Calculate the initial acceleration of A.

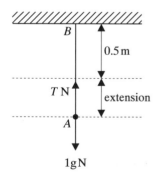

The extension of the string is $(0.75 - 0.5)\,\mathrm{m} = 0.25\,\mathrm{m}$.

Let the initial acceleration of A be $a\ \mathrm{m\,s^{-2}}$.

Using the equation of motion $F = ma$ gives:

$$T - 1g = a \qquad (1)$$

Using Hooke's law: $\qquad T = \dfrac{\lambda x}{l}$

Substituting $\lambda = 25\,\text{N}$, $x = (0.75 - 0.5)\,\text{m} = 0.25\,\text{m}$, $l = 0.5\,\text{m}$ gives:

$$T = 25 \times \frac{0.25}{0.5}$$
$$T = 12.5$$

Substituting $T = 12.5$ in equation (1) gives:

$$12.5 - \text{g} = a$$
$$a = 12.5 - 9.8 = 2.7$$

The initial acceleration of the particle is $2.7\,\text{m s}^{-2}$.

Exercise 2C

Whenever a numerical value of g is required, take $\text{g} = 9.8\,\text{m s}^{-2}$.

1 An elastic string has natural length 2 m and modulus 30 N. Find the tension in the string when the extension is 0.5 m.

2 An elastic string has modulus 25 N and the tension in the string is 40 N. Find the natural length if the extension is 0.1 m.

3 The tension in an elastic string of modulus 30 N is 50 N. The length of the string is 1.5 m. Calculate the natural length of the string.

4 A spring of modulus 40 N and natural length 0.5 m is compressed to a length of 0.4 m. Calculate the thrust produced.

5 A spring of natural length 1 m exerts a thrust of 45 N when compressed to a length of 0.75 m. Calculate the modulus of the spring.

6 A force of 5 N is applied to an elastic string of natural length 2.5 m and modulus 4 N. Calculate the extension produced.

7 An elastic spring of modulus 50 N has one end fixed. When a particle of mass 2 kg is hanging in equilibrium attached to the free end of the spring, the spring is extended by 5 cm. Find the natural length of the spring.

8 An elastic string of modulus 20 N has one end fixed. When a particle of mass 1 kg is hanging in equilibrium from the free end of the string, the length of the string is 1.4 m. Determine the natural length of the string.

The particle of mass 1 kg is removed and replaced by a particle of mass 0.75 kg. Calculate the new length of the string.

9 A particle of mass 6 kg hangs vertically from the free end A of an elastic string AB. End B of the string is fixed. If the extension in the string is equal to its natural length, determine the modulus of the string.

10 Two elastic strings, AB of modulus 15 N and natural length 1 m and CD of modulus 18 N and natural length 0.5 m are joined, B to C, to form one long string. End A is fixed to a ceiling. A particle of mass 0.5 kg is attached to D and hangs in equilibrium. Calculate the length of the combined string.

11 Two elastic strings both of natural length 1.5 m and moduli 8 N and 6 N respectively are joined together at both ends. One end of the combined string is fixed and a particle of mass 2 kg is attached to the other end and hangs in equilibrium with the string vertical. Calculate the tension in each string.

12 A ball of mass 0.5 kg is attached to one end of a cord of unstretched length 0.6 m whose other end is fixed. When a horizontal force of magnitude Q N is applied to the ball, holding the ball in equilibrium, the cord increases in length by 10% and is inclined at an angle of arcsin $\frac{3}{5}$ to the downward vertical. Choose suitable models for the ball and cord stating clearly the assumptions you make. Hence calculate the value of Q and the modulus of the cord.

13 A particle of mass 2 kg is attached to one end of a light elastic string of natural length 0.5 m and modulus 60 N whose other end is fixed. When the particle is pulled aside by a horizontal force S and held in equilibrium the extension in the string is 0.2 m. Calculate the angle the string makes with the downward vertical and the magnitude of S.

14 A particle of mass 4 kg is attached to one end A of a light elastic string AB of modulus 20 N and natural length 0.8 m. The end B of the string is attached to a point on a smooth plane inclined at an angle arcsin $\frac{5}{13}$ to the horizontal. The particle rests in equilibrium on the plane with AB along a line of greatest slope. Calculate (a) the tension in the string (b) the extension of the string.

15 A metal sphere of mass 3 kg is attached to one end A of a spring AB of modulus 90 N and natural length 0.5 m. End B of the spring is fixed to a point on a surface which is inclined at an angle arcsin $\frac{3}{5}$ to the horizontal. The sphere rests in equilibrium on the surface with AB along a line of greatest slope and A below B. A horizontal force of magnitude Q N is applied to the sphere causing the spring to be compressed to a length 0.4 m.
(a) Assuming that the surface can be modelled as a smooth plane, choose suitable models for the sphere and spring and hence calculate the value of Q.
(b) Refining your model of the surface to a rough plane and given that the sphere is on the point of moving down the plane when $Q = 35$, calculate the coefficient of friction between the surface and the sphere.

16 A particle of mass 2 kg is attached to one end A of a light spring AB of modulus 60 N and natural length 0.5 m. The spring rests on a smooth horizontal table with one end fixed. The particle is held at rest with AB in a straight line. The particle is then released. Calculate its initial acceleration (a) if AB is initially 0.7 m (b) if AB is initially 0.3 m.

17 A particle P of mass m is suspended from two strings AP and BP where A and B are attached to two points on a horizontal ceiling distance $\frac{5l}{4}$ apart as shown in the diagram.
AP is inelastic and of length $\frac{3l}{4}$. BP is elastic and of length l. The modulus of elasticity of BP is λ. Show that the natural length L of BP is given by:

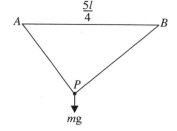

$$L = \frac{5\lambda l}{3mg + 5\lambda}$$

18 A particle of mass 0.5 kg is attached to one end A of a light elastic string AB of modulus 40 N and natural length 0.8 m. The other end B of the string is fixed and A is initially hanging in

equilibrium with the string vertical. A is then pulled vertically downwards a further 0.1 m and released. Calculate the initial acceleration of the particle.

2.4 Simple harmonic motion

We now consider a very special and important kind of motion of a particle P moving on a straight line, called **simple harmonic motion** or **S.H.M.** In this motion the acceleration is always *towards* a fixed point O of the line and is proportional to the displacement of P from O. The fixed point O is often called the **centre of oscillation**. To ensure that the acceleration of the particle is always towards O, this acceleration is usually written as:

$$\ddot{x} = -\omega^2 x$$

where ω is a real number.

As $\ddot{x}$ is the acceleration it follows that:

$$\ddot{x} = \frac{dv}{dt}$$

but as shown in chapter 1:

$$\frac{dv}{dt} = v \frac{dv}{dx}$$

So:
$$\ddot{x} = v \frac{dv}{dx} = -\omega^2 x$$

or:
$$\frac{d}{dx}\left(\frac{1}{2} v^2\right) = -\omega^2 x$$

Integrating with respect to x gives:

$$\frac{1}{2} v^2 = \int -\omega^2 x \, dx + c$$

where c is an arbitrary constant.

So:
$$\frac{1}{2}v^2 = -\omega^2 \times \frac{1}{2}x^2 + c$$
$$v^2 = -\omega^2 x^2 + 2c$$

Since the speed of P is the modulus of v, or in other words the modulus of $\frac{dx}{dt}$, it follows that the speed of P is zero when x has its maximum or minimum value. Hence c can be determined by substituting $v = 0$ and $x = a$, where a is the maximum displacement of the particle from O.

Hence:
$$0 = -\omega^2 a^2 + 2c$$
$$2c = \omega^2 a^2$$

and so:
$$v^2 = -\omega^2 x^2 + \omega^2 a^2$$

■
$$v^2 = \omega^2 (a^2 - x^2) \qquad (1)$$

This is a standard result which should be memorised. It may be quoted and used in the examination without first proving it.

As $v = \dfrac{\mathrm{d}x}{\mathrm{d}t}$, an expression for x can be found by integrating equation (1).

From equation (1): $\qquad \dfrac{\mathrm{d}x}{\mathrm{d}t} = \pm\omega(a^2 - x^2)^{\frac{1}{2}}$

Separating the variables and integrating gives:

$$\int \frac{\mathrm{d}x}{(a^2 - x^2)^{\frac{1}{2}}} = \pm \int \omega\,\mathrm{d}t$$

For the left hand integral a substitution is required as shown in Book P2 chapter 9.

Substitute: $\qquad\qquad\qquad x = a\sin\theta$

So: $\qquad\qquad\qquad \mathrm{d}x = a\cos\theta\,\mathrm{d}\theta$

and: $\qquad\qquad (a^2 - x^2)^{\frac{1}{2}} = (a^2 - a^2\sin^2\theta)^{\frac{1}{2}}$
$$= a\cos\theta$$

since $\cos^2\theta + \sin^2\theta = 1$ (Book P2 chapter 6).

Hence:
$$\int \frac{\mathrm{d}x}{(a^2 - x^2)^{\frac{1}{2}}} = \int \frac{a\cos\theta\,\mathrm{d}\theta}{a\cos\theta}$$
$$= \int \mathrm{d}\theta$$

So:
$$\int \mathrm{d}\theta = \pm \int \omega\,\mathrm{d}t$$
$$\theta = \pm\omega t + \alpha$$

and: $\qquad\qquad\qquad \sin\theta = \sin(\omega t + \alpha)$

or: $\qquad\qquad\qquad x = a\sin(\omega t + \alpha) \qquad (2)$

This function has period $\dfrac{2\pi}{\omega}$ and maximum and minimum values $\pm a$, so $-a \leqslant x \leqslant a$. a is called the **amplitude** of the motion.

To investigate the motion further, different values of the constant of integration, α, must be considered.

Case (i) $\alpha = 0$

When $\alpha = 0$, equation (2) becomes:

$$x = a \sin \omega t$$

The graph of $x = a \sin \omega t$ is:

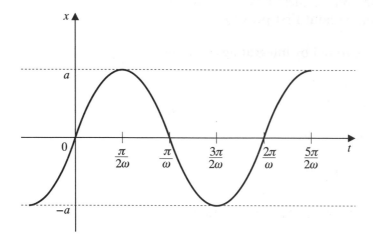

The graph shows that $x = a \sin \omega t$ gives the displacement from the centre of oscillation of a particle moving with S.H.M. of amplitude a and period $\dfrac{2\pi}{\omega}$ which is at the centre of the oscillation (that is $x = 0$) when $t = 0$.

Case (ii) $\alpha = \frac{\pi}{2}$

when $\alpha = \frac{\pi}{2}$, equation (2) becomes

$$x = a \sin\left(\omega t + \frac{\pi}{2}\right)$$

In Book P1 chapter 7 it was shown that the graph of $x = a \sin(\omega t + \frac{\pi}{2})$ is the same shape as that of $x = a \sin \omega t$ but is translated $\frac{\pi}{2}$ to the left.

Hence the graph of $x = a \sin\left(\omega t + \frac{\pi}{2}\right)$ is:

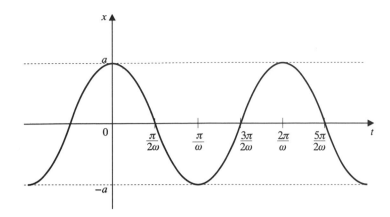

which you will recognise as being the graph of $x = a\cos\omega t$.

Hence the equation $x = a\cos\omega t$ gives the displacement from the centre of oscillation of a particle moving with S.H.M. of amplitude a and period $\dfrac{2\pi}{\omega}$ which has x maximum when $t = 0$.

Case (iii) α is neither of the above.

When α takes some value other than 0 or $\frac{\pi}{2}$ the graph of $x = a\sin(\omega t + \alpha)$ is a translation of the graph of $x = a\sin\omega t$ through a distance α to the left.

The graph of $x = a\sin(\omega t + \alpha)$ is therefore:

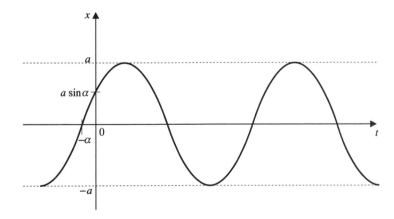

$x = a\sin(\omega t + \alpha)$ gives the solution when the particle is neither at the centre nor at an extreme point of the oscillation when $t = 0$.

Thus for S.H.M. of amplitude a defined by the equation $\ddot{x} = -\omega^2 x$:

$$v^2 = \omega^2(a^2 - x^2)$$

and: $\quad x = a \sin \omega t \qquad$ if $x = 0$ when $t = 0$

$\qquad x = a \cos \omega t \qquad$ if $x = a$ when $t = 0$

$\qquad x = a \sin(\omega t + \alpha) \quad$ if x is at some other point when $t = 0$.

Example 11

A particle P is moving along a straight line with S.H.M. of amplitude 0.5 m. It passes through the centre of the oscillation O with speed $5\,\text{m s}^{-1}$. Calculate the period of the oscillation and the speed of the particle when $OP = 0.25\,\text{m}$.

Using: $$v^2 = \omega^2(a^2 - x^2)$$

with $a = 0.5\,\text{m}$, $v = 5\,\text{m s}^{-1}$ and $x = 0\,\text{m}$ gives:

$$5^2 = \omega^2(0.5^2 - 0)$$

$$\omega^2 = \frac{5^2}{0.5^2}$$

$$\omega = \frac{5}{0.5} = 10$$

Using: $$\text{period} = \frac{2\pi}{\omega}$$

gives: $$\text{period} = \frac{2\pi}{10} = \frac{\pi}{5}$$

The period of oscillation is $\dfrac{\pi}{5}$ seconds.

Using: $$v^2 = \omega^2(a^2 - x^2)$$

with $a = 0.5\,\text{m}$, $x = 0.25\,\text{m}$ and $\omega = 10$ gives:

$$v^2 = 10^2(0.5^2 - 0.25^2)$$

$$v = 10\sqrt{(0.5^2 - 0.25^2)}$$

$$v = 4.33\dot{}$$

The speed of P when $OP = 0.25\,\text{m}$ is $4.33\,\text{m s}^{-1}$.

Example 12

A particle is moving in a straight line with S.H.M. Its maximum speed is $5\,\text{m s}^{-1}$ and its maximum acceleration is $12\,\text{m s}^{-2}$. Calculate the amplitude and period of the oscillation.

The maximum speed occurs when the acceleration is zero. Since $\ddot{x} = -\omega^2 x$ it occurs when $x = 0$. The acceleration has its maximum magnitude when x has its maximum magnitude. Hence it occurs when $x = \pm a$.

Using: $$v^2 = \omega^2(a^2 - x^2)$$

with $v = 5\,\mathrm{m\,s^{-1}}$ and $x = 0$ gives:

$$5^2 = \omega^2 a^2 \qquad (1)$$

Using: $$\ddot{x} = -\omega^2 x$$

gives: $$|\ddot{x}| = |-\omega^2 x|$$

And so: $$|\ddot{x}| = \omega^2 |x|$$

As the maximum magnitude for $\ddot{x}$ is $12\,\mathrm{m\,s^{-2}}$ and this occurs when $x = \pm a$, it follows that

$$12 = \omega^2 a \qquad (2)$$

Dividing (1) by (2) gives:

$$\frac{5^2}{12} = \frac{\omega^2 a^2}{\omega^2 a}$$

$$a = \frac{5^2}{12} = 2.083$$

The amplitude is $2.08\,\mathrm{m}$.

Substituting $a = \dfrac{25}{12}$ in equation (2) gives:

$$\frac{25}{12}\,\omega^2 = 12$$

$$\omega^2 = \frac{12^2}{25}$$

$$\omega = \frac{12}{5} = 2.4$$

Hence: $$\text{period} = \frac{2\pi}{\omega} = \frac{2\pi}{2.4} = 2.617$$

The period is $2.62\,\mathrm{s}$.

There are many real-life situations that involve oscillatory motion. Examples include the motion of the pistons in an engine, the swinging pendulum of a clock and the motion of an object bobbing up and down on the waves. A particle moving with S.H.M. can provide a good model for the mathematical analysis of such motions.

Example 13

A buoy on the surface of the sea is tied to a pier and oscillates in a vertical line with S.H.M. The period of oscillation is $\dfrac{3\pi}{4}$ seconds and the maximum speed of the buoy is $4\,\mathrm{m\,s^{-1}}$. At the highest point of the motion the buoy is $5\,\mathrm{m}$ below the pier. Choose a suitable model for the buoy and hence calculate the amplitude of the motion and find the time taken by the buoy to fall from its highest point to 6 m below the pier.

The buoy will be small compared with other measurements and so can be modelled as a particle.

Using:

$$\text{period} = \frac{2\pi}{\omega}$$

with period $= \frac{3\pi}{4}$ gives:

$$\frac{3\pi}{4} = \frac{2\pi}{\omega}$$

$$\omega = \frac{8}{3}$$

Using:

$$v^2 = \omega^2(a^2 - x^2)$$

with $\omega = \frac{8}{3}$, $v = 4$, $x = 0$ gives:

$$4^2 = \frac{8^2}{3^2} \times a^2$$

$$a^2 = \frac{4^2 \times 3^2}{8^2}$$

$$a = \frac{4 \times 3}{8} = 1.5$$

The amplitude is 1.5 m.

Taking $t = 0$ at the highest point, the form required is $x = a\cos\omega t$.

At the highest point, the buoy is 5 m below the pier. To find the time when the buoy is 6 m below the pier you need to find the time when the buoy has fallen a distance $(6 - 5)\,\text{m} = 1\,\text{m}$.

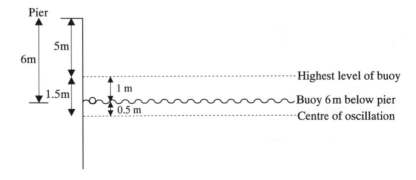

So when the buoy has fallen 1 m, $x = 1.5 - 1 = 0.5$ m.

Using:
$$x = a \cos \omega t$$

with $a = 1.5$ m, $\omega = \frac{8}{3}$ and $x = 0.5$ m gives:

$$0.5 = 1.5 \cos \tfrac{8}{3} t$$
$$\tfrac{1}{3} = \cos \tfrac{8}{3} t$$
$$\tfrac{8}{3} t = \arccos \left(\tfrac{1}{3} \right)$$
$$t = \tfrac{3}{8} \arccos \left(\tfrac{1}{3} \right)$$
$$t = 0.4616$$

The time taken is 0.462 s.

A geometrical approach to simple harmonic motion

Consider a particle P which is moving round a circle of radius a, centre the origin O, with a constant angular speed ω in an anticlockwise sense. Let A be the foot of the perpendicular from P onto the x-axis.

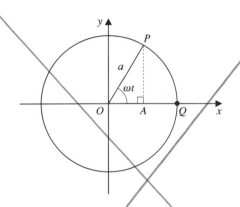

If P is at the point Q when $t = 0$ then at time t seconds:
$$\angle POA = \omega t$$

and so:
$$OA = x = a \cos \omega t$$

Differentiating with respect to time gives:
$$\dot{x} = -a\omega \sin \omega t$$

and:
$$\ddot{x} = -a\omega^2 \cos \omega t$$

So:
$$\ddot{x} = -\omega^2 x$$

The point A is therefore moving along the x-axis with S.H.M. of amplitude a and period $\dfrac{2\pi}{\omega}$.

This connection between circular motion and S.H.M. can be particularly useful when calculating the time for a particle to move between two points of the oscillation. The circle associated with any particular S.H.M. is often called the **reference circle**.

Example 14

A particle P is moving along a straight line with S.H.M. centre O with a period 5 s and amplitude 0.4 m. A and B are points on the line of P's motion such that OA is 0.3 m and OB is 0.2 m with A and B on opposite sides of O. Calculate the time taken by P to move directly from A to B.

First draw a diagram showing a circle of radius 0.4 m with a diameter representing the line on which P is moving. The points A and B are on this diameter on opposite sides of O as shown:

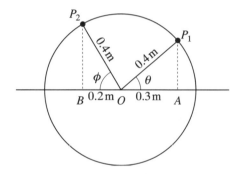

The time taken by P to move directly from A to B is the time taken by a particle moving on the reference circle with angular speed ω to move from P_1 to P_2; that is, the time to move round an arc of the circle which subtends an angle of $(\pi - \theta - \phi)^c$ at O.

From the diagram: $\qquad \theta = \arccos\left(\dfrac{0.3}{0.4}\right) = \arccos\left(\dfrac{3}{4}\right)$

$$\phi = \arccos\left(\dfrac{0.2}{0.4}\right) = \arccos\left(\dfrac{1}{2}\right)$$

So the required angle is:

$$\pi - \theta - \phi = \pi - \arccos\left(\tfrac{3}{4}\right) - \arccos\left(\tfrac{1}{2}\right)$$

The particle on the reference circle takes 5 s (the period) to travel a distance corresponding to an angle of 2π. Let the time taken to travel $\pi - \theta - \phi$ be T s.

Since the angular speed is constant:

$$\frac{T}{\pi - \theta - \phi} = \frac{5}{2\pi}$$

$$T = \frac{5}{2\pi} \times \left(\pi - \arccos\left(\tfrac{3}{4}\right) - \arccos\left(\tfrac{1}{2}\right) \right)$$

$$T = 1.091$$

The time to travel from A to B is $1.09\,\text{s}$.

Exercise 2D

1 A particle P is moving on a straight line with S.H.M. of amplitude $0.3\,\text{m}$. It passes through the centre of the oscillation O with speed $2\,\text{m\,s}^{-1}$. Calculate the period of oscillation and the speed of the particle when $OP = 0.1\,\text{m}$.

2 A particle P is moving on a straight line with S.H.M. of amplitude $0.5\,\text{m}$ and period $\dfrac{2\pi}{5}\,\text{s}$. Calculate the maximum speed of the particle.

3 A particle P is moving on a straight line with S.H.M. of period $\dfrac{\pi}{3}\,\text{s}$. Its maximum speed is $5\,\text{m\,s}^{-1}$. Calculate the amplitude of the motion and the speed of P $0.2\,\text{s}$ after passing through the centre of oscillation.

4 A particle is moving on a straight line with S.H.M. Its maximum speed is $10\,\text{m\,s}^{-1}$ and its maximum acceleration is $10\,\text{m\,s}^{-2}$. Calculate the period and amplitude of the motion.

5 A particle is moving on a straight line with S.H.M. Its maximum speed is $12\,\text{m\,s}^{-1}$ and its maximum acceleration is $10\,\text{m\,s}^{-2}$. Calculate the speed of the particle when it is $2\,\text{m}$ from the centre of oscillation.

6 A particle is moving in a straight line with S.H.M. of period $2\pi\,\text{s}$ about a centre O. When the particle is $0.5\,\text{m}$ from O its speed is zero. Calculate the maximum speed of the particle and its speed when it is $0.25\,\text{m}$ from O.

7 A particle P is moving in a straight line with S.H.M. of period $\pi\,\text{s}$ about a centre O. When $OP = 2\,\text{m}$ P has a speed of $3\,\text{m\,s}^{-1}$. Find the amplitude of the motion.

Given that when $t = 0$ P is passing through O, find an expression for the displacement x m of P from O at any subsequent time t s.

8 A particle P is moving in a straight line with S.H.M. centre O. P passes through O with speed $6\,\text{m s}^{-1}$ and performs 4 complete oscillations per second. Calculate the amplitude of the motion and the distance OP when the magnitude of P's acceleration is $30\,\text{m s}^{-2}$.

9 A particle P is moving on a straight line with S.H.M. between two points A and B. O is the mid-point of AB. When $OP = 1$ m, P has speed $4\,\text{m s}^{-1}$. When $OP = 0.5$ m P has speed $6\,\text{m s}^{-1}$. Determine the distance AB and the period of the motion.

10 A piston of mass 1.5 kg is moving with S.H.M. inside a cylinder. It performs 20 complete oscillations per minute and the distance between the extreme points of its motion is 2 m. Calculate the maximum value of the kinetic energy possessed by the piston during its motion, stating clearly any assumptions you have made about the piston in order to calculate this.

11 A marker buoy is oscillating in a vertical line with S.H.M. The buoy takes 3 s for one oscillation and is seen to rise and fall through a distance of 0.75 m. Choose a suitable model for the marker buoy and hence calculate its maximum speed and the time taken by the buoy to fall a distance of 0.5 m from its highest point.

12 A particle is moving along the x-axis so that its displacement x m from O at time t s is given by $x = 5\sin 3t$. Prove that the particle is moving with S.H.M. and state the amplitude and period of the motion. Calculate (a) the maximum speed of the particle (b) the value of t the first time its speed is $10\,\text{m s}^{-1}$ (c) the value of t the first time $x = 2.5$.

13 A particle P is moving along the x-axis so that its displacement x m from O at time t s is given by $x = 2\sin(5t + \frac{1}{2})$. Prove that the particle is moving with S.H.M. and state the amplitude and period of the motion. Determine the value of x when $t = 0$ and the time when P passes through O for the first time.

14 Points A, O, B lie in that order on a straight line. A particle P is moving on the line with S.H.M. of period 3 s, amplitude 0.6 m

and centre O. Given that OA is 0.3 m and OB is 0.4 m, calculate the time taken by P to move directly from A to B.

15 Points O, A, B lie in that order on a straight line. A particle P is moving on the line with S.H.M. of period 4 s, amplitude 0.5 m and centre O. OA is 0.1 m and OB is 0.3 m. When $t = 0$, P passes through B travelling in the direction OB. Calculate the time when P reaches A.

16 The water level in a harbour can be assumed to rise and fall with simple harmonic motion. On a certain day, low tide will occur at 12 noon and the depth of the water will be 3 m. The subsequent high tide will occur at 6.20 pm and the water will then be 15 m deep. A ship which needs 6 m depth of water wishes to use the harbour. Find, to the nearest minute, the earliest time it can enter the harbour on this day and the time by which it must leave.

2.5 The simple pendulum

A **simple pendulum** consists of a particle, sometimes called a bob, attached to one end of a light string which is fixed at the other end and swings through small angles on either side of the vertical.

Let the length of the string be l.

Consider the situation when the particle is at a distance s from the centre O of the oscillation, where s is measured along the arc that the particle describes.

Then: $$s = l\theta$$

Differentiating with respect to time gives:
$$\dot{s} = l\dot{\theta}$$
and: $$\ddot{s} = l\ddot{\theta}$$

The component of the force acting on the particle in the direction of the tangent to the arc described is $mg\sin\theta$ *towards* the centre of oscillation, O. So by Newton's second law:
$$F = ma$$
$$mg\sin\theta = -m\ddot{s}$$
or: $$mg\sin\theta = -ml\ddot{\theta}$$

But for small angles: $$\sin\theta \simeq \theta$$

(This is shown in Book P3. Alternatively use your calculator to convince yourself of the validity of this statement.)

And so:
$$mg\theta = -ml\ddot{\theta}$$

$$\ddot{\theta} = -\frac{g}{l}\theta$$

Comparing this equation with the standard S.H.M. equation $\ddot{x} = -\omega^2 x$ shows that the pendulum is moving with S.H.M. of period $2\pi\sqrt{\left(\dfrac{l}{g}\right)}$.

The calculation above was made using several assumptions. The bob was assumed to be a particle, the string was assumed to be light and inelastic, the point of support was assumed to be fixed and the amplitude of the oscillation was small so that $\sin\theta \simeq \theta$. A simple pendulum model is valid for a body swinging on the end of a string or rod which is fixed at the other end only if these assumptions are reasonable. In particular:

(i) The mass of the body must be large compared with the weight of the string or rod, otherwise the string cannot be assumed light.

(ii) The size of the body must be small compared with the length of the string or the body cannot be assumed to be a particle.

(iii) The amplitude of the oscillation must be small. The formula has been found to be very accurate for amplitudes up to 0.26^c (approximately $15°$) but thereafter the accuracy diminishes.

The investigation of the motion of a pendulum which does not satisfy these conditions is beyond the scope of this book. However, provided these assumptions are valid for a particular pendulum arrangement then the simple pendulum model is valid. The period of oscillation, T, is then independent of the mass of the bob and is given by:

■
$$T = 2\pi\sqrt{\left(\frac{l}{g}\right)}$$

where l is the length of the string or rod.

Example 15
A simple pendulum is found to have a period of 2 seconds. Calculate the length of the string.

$$\text{Period} = 2\pi\sqrt{\left(\frac{l}{g}\right)}$$

So:
$$2 = 2\pi\sqrt{\left(\frac{l}{9.8}\right)}$$

$$\sqrt{\left(\frac{l}{9.8}\right)} = \frac{2}{2\pi} = \frac{1}{\pi}$$

Squaring both sides:
$$\frac{l}{9.8} = \frac{1}{\pi^2}$$

$$l = \frac{9.8}{\pi^2} = 0.9929$$

The length of the pendulum is 0.993 m.

Exercise 2E

Whenever a numerical value of g is required, take g $= 9.8\,\mathrm{m\,s^{-2}}$.

1 A simple pendulum is performing small oscillations. Calculate the period of the pendulum given that the length is (a) 0.5 m (b) 1.3 m (c) 70 cm.

2 A simple pendulum is performing small oscillations. Calculate the length of the pendulum given that the period is

(a) $\frac{3\pi}{8}$s (b) $\frac{4\pi}{9}$s (c) 1 s.

3 A simple pendulum takes one second to perform half a complete oscillation. Calculate the length of string required for the pendulum. (A pendulum with this period is called a seconds pendulum.)

4 A simple pendulum has length L and period T. If the length is reduced to $\frac{1}{2}L$, calculate the new period in terms of T.

5 A simple pendulum has period T and length L. The period is to be increased to $3T$. Calculate the new length in terms of L.

6 A plumb line consists of a light cord of length 2 m with a metal bob attached to one end. The other end of the cord is fixed and the bob is hanging freely. The bob is slightly displaced and the plumb line is seen to oscillate. State the assumptions you would be making about the cord, the metal bob and the oscillations if you were to model the plumb line as a simple pendulum. Assuming the simple pendulum model to be a valid one, calculate the period of the oscillations.

7 A boy is swinging on the end of a rope. The other end of the rope is tied to a branch of a tree. What assumptions must be made if his motion is to be considered to be that of a simple pendulum? Given that this model is valid and the period of oscillation is 3.5 s, calculate the length of the rope.

2.6 Horizontal oscillations of a particle attached to the end of an elastic string or spring

Consider a particle P resting on a smooth horizontal surface and attached to one end of an elastic spring whose other end is fixed to a point O on the surface.

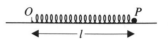

When there is no horizontal force applied to P the spring will be unstretched and so at its natural length l. However, if P is pulled aside in the direction OP and then released the spring will be stretched and so will exert a tension on P which will result in an acceleration.

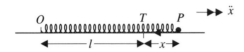

Let the extension of the spring be x.

By Hooke's law the tension T in the spring is given by:

$$T = \frac{\lambda x}{l}$$

where l is the natural length and λ the modulus of the spring.

The acceleration $\ddot{x}$ of P will be directed *away* from the fixed point O of the spring as that is the direction of *increasing* x.

By Newton's second law, $F = ma$ and so:

$$-T = m\ddot{x}$$

where m is the mass of P.

Hence:
$$m\ddot{x} = -\frac{\lambda x}{l}$$

or:
$$\ddot{x} = -\frac{\lambda x}{ml} \qquad (1)$$

But λ, m and l all have positive, constant values. So equation (1) is of the form $\ddot{x} = -\omega^2 x$ and hence the particle is moving with S.H.M.

where $\omega^2 = \dfrac{\lambda}{ml}$

Since:
$$\text{period} = \frac{2\pi}{\omega}$$

the period of the particle's oscillations is $2\pi\sqrt{\left(\dfrac{ml}{\lambda}\right)}$.

The initial extension is also the maximum value of x. So the amplitude of the motion is the same as the initial extension.

The above work used Hooke's law to determine the tension in the spring. A *spring* will always have a tension or compression and so a particle attached to the end of a spring will perform complete oscillations whether it is pulled aside (spring initially stretched) or pushed aside (spring initially compressed) before release. However, for a particle attached to the end of a *string*, the motion will be oscillatory only while the string is taut. Once the string becomes slack there will be no tension and the particle will continue to move with constant speed until the string becomes taut again.

In the above diagram the string is stretched and P will accelerate.

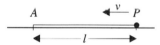

Once the string returns to its natural length there is no tension and hence P has no acceleration but it has already acquired a velocity.

When P has moved a distance equal to twice the natural length of the string, the string will become taut again. The resulting tension will cause P to decelerate and oscillatory motion will once more take place.

Example 16

A small ball of mass 0.5 kg rests on a smooth horizontal table attached to one end of a piece of elastic of natural length 0.8 m and modulus 20 N. The other end of the elastic is fixed to a point on the table. The ball is pulled aside so that the elastic is extended by 0.2 m and then released.

(a) Describe the models you would use for the ball and elastic when investigating the resulting motion.
(b) Determine the speed of the ball when the elastic returns to its natural length.

    (c) Calculate the time that elapses between the string becoming slack and becoming taut again. Hence calculate the time taken by the ball to return to its starting point for the first time.

    (d) Describe briefly how your model could be refined.

(a) The ball is small compared with the length of the elastic and so can be modelled as a particle. The elastic can be modelled as a light elastic string.

(b) Consider the ball when the elastic is stretched x m. The total length of the elastic is then $(0.8 + x)$ m. Let the tension in the elastic be T N and the acceleration of the ball be $\ddot{x}$ m s^{-2} as shown.

Using Hooke's law $T = \dfrac{\lambda x}{l}$ with $\lambda = 20$ N and $l = 0.8$ m gives:

$$T = \frac{20x}{0.8}$$
$$T = 25x \qquad\qquad (1)$$

Using Newton's second law $F = ma$ with $m = 0.5$ kg gives:

$$T = -0.5\ddot{x} \qquad\qquad (2)$$

Combining equations (1) and (2) gives:

$$25x = -0.5\ddot{x}$$

So:

$$\ddot{x} = -50x$$

Comparing this with the standard S.H.M. equation $\ddot{x} = -\omega^2 x$ shows $\omega = \sqrt{50}$.

The amplitude of the motion is the same as the initial extension of the string. So the amplitude is 0.2 m.

To determine the speed of the ball when the string becomes slack you must use the standard S.H.M. equation:

$$v^2 = \omega^2(a^2 - x^2)$$

with $\omega = \sqrt{50}$, $a = 0.2$ and $x = 0$.

So:

$$v^2 = 50(0.2^2 - 0)$$
$$v = 1.414$$

The speed of the ball when the elastic returns to its natural length is 1.41 m s^{-1}.

(c)

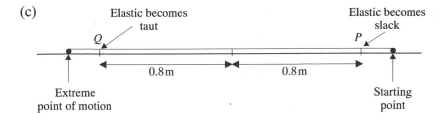

The ball will move from P to Q while the string is slack. As the natural length of the string is 0.8 m, distance $PQ = 2 \times 0.8$ m $= 1.6$ m

The speed of the ball is $1.414 \, \text{m s}^{-1}$.

As: $$\text{time} = \frac{\text{distance}}{\text{speed}}$$

the time is $\left(\dfrac{1.6}{1.414}\right)$ s $= 1.131$ s

The string is slack for 1.13 s.

To calculate the time taken by the ball to return to its starting point you must include the time for which the ball is oscillating. From (b) above $\omega = \sqrt{50}$

So the period of the oscillation $= \dfrac{2\pi}{\omega} = \dfrac{2\pi}{\sqrt{50}}$

The whole motion consists of half an oscillation at each end of the ball's path and the constant speed section in the middle. The middle section is covered twice.

So: total time = period of oscillation $+ 2 \times$ time from P to Q

$$= \frac{2\pi}{\sqrt{50}} + 2 \times 1.131$$

$$= 3.150$$

The particle will return to its starting point after 3.15 s.

(d) The table is unlikely to be perfectly smooth. The model could therefore be refined by including the frictional force between the table and the ball.

Exercise 2F

1 A particle P of mass 0.2 kg is attached by an elastic string of modulus 15 N and natural length 1 m to a point A of the smooth horizontal surface on which P rests. P receives an impulse of

magnitude $0.5\,\mathrm{N\,s}$ in the direction AP. Show that while the string is taut the motion of P is simple harmonic and calculate the period and amplitude of the motion.

2 A particle P of mass $1.2\,\mathrm{kg}$ is attached by an elastic spring of modulus $100\,\mathrm{N}$ and natural length $0.5\,\mathrm{m}$ to a point A on the smooth horizontal table on which P rests. P receives an impulse of magnitude $2\,\mathrm{N\,s}$ in the direction AP. Given that t seconds after the blow the displacement of P from its initial position is $x\,\mathrm{m}$, find an equation for x in terms of t and calculate the maximum acceleration of P.

3 A small ball of mass 0.7 kg is attached to one end of a light elastic spring of natural length $0.6\,\mathrm{m}$ and modulus $84\,\mathrm{N}$. The other end of the spring is fixed to a point on the smooth horizontal surface on which the ball rests. The ball is pushed aside so that the spring has length $0.4\,\mathrm{m}$ and then released. Calculate the period of the resulting oscillations and the maximum speed of the ball.

4 A small ball of mass $0.3\,\mathrm{kg}$ rests on a smooth horizontal table attached to one end of a piece of elastic of natural length $1\,\mathrm{m}$ and modulus $48\,\mathrm{N}$. The other end of the elastic is fixed to a point on the table. The ball is pulled aside so that the elastic measures $1.3\,\mathrm{m}$ and then released.
(a) Choose suitable models for the ball, elastic and table so that you can investigate the motion of the ball.
(b) Determine the speed of the ball when the elastic is slack.
(c) Calculate the time taken by the ball to return to its starting point for the first time.
(d) Your models allow a force (other than gravity) to be neglected. Explain briefly why this is possible. If this force were included, how would your answer to part (b) be affected?

5 A particle of mass $0.75\,\mathrm{kg}$ rests on a smooth horizontal surface attached to one end of a light elastic string of natural length $1.5\,\mathrm{m}$ and modulus $20\,\mathrm{N}$. The other end of the string is attached to a point on the surface. The particle is pulled aside so that the string measures $2\,\mathrm{m}$ and released. State the amplitude of the resulting oscillation and calculate the speed of the particle when the string becomes slack.

Before the string becomes taut again the particle hits a vertical surface which is at right angles to the particle's direction of motion. The particle rebounds with speed half that with which it hits the surface. Calculate the amplitude of the oscillation which takes place when the string is once more taut.

6 A piston of mass 2 kg is moving in a smooth cylinder which is fixed with its axis horizontal. The piston is attached to the end of the cylinder by a spring of natural length 40 cm and modulus 500 N. The piston is pushed aside until the spring is compressed to a length of 30 cm and then released.

(a) Choose suitable models for the spring and piston stating clearly any assumptions you make.

(b) Determine the period of the resulting oscillations and the maximum speed of the piston.

(c) Describe briefly how your model could be refined.

7 A small ball of mass 1.2 kg is attached to an elastic string of unstretched length 0.75 m and modulus 60 N. The other end of the string is fixed to a point of the smooth horizontal floor on which the ball rests. The ball is pulled aside until the string measures 0.9 m and released. Calculate the speed of the ball when the string becomes slack.

When the ball has travelled a distance 0.6 m from its point of release, the floor becomes rough. Given that the coefficient of friction between the ball and the floor is 0.3, determine the time for which the ball is moving.

8 Two identical elastic strings of natural length 0.8 m and modulus 4 N are attached to a particle P of mass 0.75 kg. Their free ends are attached to two points A and B on a smooth horizontal surface where $AB = 3$ m. P is held at a point C, where $AC = 1$ m and A, C and B are collinear, and released. Show that the subsequent motion is simple harmonic and give the amplitude and period of the motion. Calculate the maximum speed of P and hence its maximum kinetic energy.

Vertical oscillations of a particle attached to the end of an elastic string or spring

Consider a particle which is hanging in equilibrium attached to the end of an elastic string or spring, the other end of which is fixed. If the particle is pulled downwards and released it will oscillate in a vertical line about its equilibrium position.

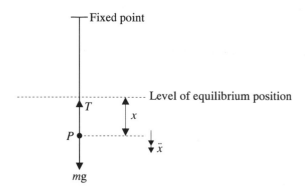

Let the particle be a distance x below its equilibrium position. The acceleration of P will be $\ddot{x}$ *away* from the equilibrium position as that is the direction of *increasing* x. By Hooke's law the tension in the string or spring is given by $T = \dfrac{\lambda}{l} \times$ extension where l is the natural length of the string or spring and λ is the modulus. Let the extension in the string or spring when the particle is in equilibrium be e. Then the total extension in the string or spring is $x + e$.

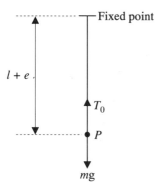

Let the tension when the particle is in equilibrium be T_0. In the equilibrium position resolving vertically gives:

$$T_0 = mg$$

Hooke's law gives:
$$T_0 = \frac{\lambda e}{l}$$

And so:
$$\frac{\lambda e}{l} = mg$$

$$e = \frac{mgl}{\lambda}$$

Hence the total extension when the particle is displaced a distance x from the equilibrium position is $e + x = \frac{mgl}{\lambda} + x$

The tension, T, at this point is given by:

$$T = \frac{\lambda}{l}\left(\frac{mgl}{\lambda} + x\right) = mg + \frac{\lambda x}{l}$$

Using Newton's second law, $F = ma$, gives:

$$mg - T = m\ddot{x}$$

and so:
$$mg - \left(mg + \frac{\lambda x}{l}\right) = m\ddot{x}$$

$$m\ddot{x} = -\frac{\lambda x}{l}$$

$$\ddot{x} = -\frac{\lambda x}{ml}$$

As in the case with horizontal oscillations, the particle is moving with S.H.M. of period $2\pi\sqrt{\left(\frac{ml}{\lambda}\right)}$.

As in the horizontal case, a particle attached to the end of a spring will perform complete oscillations. In the case of a string however, the S.H.M. equation will only hold while the string is taut. That is, the particle must not rise above its equilibrium position a distance greater than its equilibrium extension e. The greatest distance that can be reached above the equilibrium position while moving with S.H.M. is the amplitude. So the amplitude must be less than or equal to the equilibrium extension e. If this is not the case the string will become slack when $x = -e$ and the particle will then move freely under gravity until it falls back to the position where $x = -e$ once more.

To summarise:

A particle which is hanging in equilibrium attached to the end of an elastic *spring* and is displaced vertically from its equilibrium position and then released will move with S.H.M. performing complete oscillations.

A particle attached to the end of an elastic *string* and similarly displaced and released will move with S.H.M. performing complete oscillations if the initial displacement is less than the equilibrium extension e. When the string becomes slack the particle will move freely under gravity.

Example 17

A piston of mass $1.5\,\text{kg}$ moves in a smooth cylinder which is fixed with its axis vertical. It is attached to the base of the cylinder by a spring of natural length $20\,\text{cm}$ and modulus $200\,\text{N}$ as shown in the diagram.

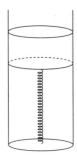

(a) Choose suitable models for the spring and piston, stating clearly any assumptions you make.
(b) Determine the compression of the spring when the piston rests in equilibrium.

The piston is now pushed downwards a further $0.5\,\text{cm}$ and released.

(c) Calculate the period of the resulting oscillations and the maximum speed of the piston.

(a) The mass of the spring can be assumed to be small compared with that of the piston. The spring can therefore be modelled as a light elastic spring.

The piston is attached to the end of the spring. Assuming the piston is a uniform solid and the spring is attached to the centre of the base of the piston, the weight of the piston will act along the line of the spring. Thus the piston will have the same effect on the spring as would a particle of mass equal to that of the piston attached to the end of the spring. In other words, the piston can be modelled as a particle.

(b)

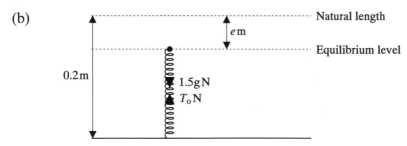

Consider the piston when it is in equilibrium. A diagram of the model is shown above.

Let the thrust in the spring be $T_0\,\text{N}$ and the equilibrium compression be $e\,\text{m}$.

Resolving vertically gives:

$$T_0 = 1.5\,g$$

Using Hooke's law:

$$T = \frac{\lambda \times \text{extension}}{\text{natural length}}$$

with $T = 1.5\,g\,N$, $\lambda = 200\,N$, extension e m and natural length $20\,cm = 0.2\,m$ gives:

$$1.5\,g = \frac{200e}{0.2}$$

$$e = \frac{1.5 \times 9.8 \times 0.2}{200}$$

$$e = 0.0147$$

The compression when the piston is in equilibrium is $0.0147\,m$ or $1.47\,cm$.

(c) Consider the piston when it is a distance x below its equilibrium position. The acceleration will be $\ddot{x}\,m\,s^{-2}$ downwards and the compression of the spring will be $(0.0147 + x)\,m$.

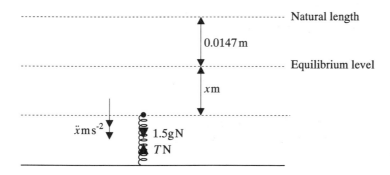

Using Newton's second law $F = ma$ with $m = 1.5\,kg$ gives:

$$1.5\,g - T = 1.5\ddot{x} \qquad (1)$$

Using Hooke's law:

$$T = \frac{\lambda \times \text{extension}}{\text{natural length}}$$

with $\lambda = 200\,N$, extension $= (0.0147 + x)\,m$ and natural length $0.2\,m$ gives:

$$T = \frac{200(0.0147 + x)}{0.2}$$

Substituting this expression for T in equation (1) gives:

$$1.5 \times 9.8 - \frac{200(0.0147 + x)}{0.2} = 1.5\ddot{x}$$

$$14.7 - 14.7 - 1000x = 1.5\ddot{x}$$

$$\ddot{x} = -\frac{1000}{1.5}x$$

or:

$$\ddot{x} = -\frac{2000}{3}x \qquad (2)$$

Comparing equation (2) with the standard S.H.M equation $\ddot{x} = -\omega^2 x$ gives:

$$\omega^2 = \frac{2000}{3}$$

So:

$$\omega = \sqrt{\left(\frac{2000}{3}\right)}$$

Since:

$$\text{period} = \frac{2\pi}{\omega}$$

this gives:

$$\text{period} = 2\pi\sqrt{\left(\frac{3}{2000}\right)} = 0.2433$$

The period of the resulting oscillations is 0.243 s.

The speed of the piston is found by using the standard formula:

$$v^2 = \omega^2(a^2 - x^2)$$

The maximum speed will occur when x^2 is minimum; that is when $x = 0$.

In the equation a is the maximum distance of the piston from its equilibrium position. As the piston was pushed downwards until it was 0.5 cm below its equilibrium position, $a = 0.5$ cm $= 0.005$ m.

Using:

$$v^2 = \omega^2(a^2 - x^2)$$

with $a = 0.005$, $\omega^2 = \frac{2000}{3}$ and $x = 0$ gives:

$$v^2 = \frac{2000}{3}(0.005^2 - 0)$$

$$v = 0.1290$$

The maximum speed is 0.129 m s^{-1}.

Example 18

A particle P of mass 0.1 kg is hanging in equilibrium attached to one end of a light elastic string of natural length 0.5 m and modulus 5 N. The other end of the string is fixed. P is pulled downwards a further 0.2 m and released. Calculate the time that elapses before the string becomes slack and the further time before the string is once more taut.

First the equilibrium level must be determined. Let the extension when the particle is in equilibrium be e m and the tension in the string be T_0 N.

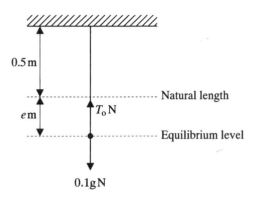

For the particle in equilibrium, resolving vertically gives:

$$T_0 = 0.1g$$

Using Hooke's law:

$$T = \frac{\lambda \times \text{extension}}{\text{natural length}}$$

with $\lambda = 5$ N and natural length 0.5 m gives:

$$T_0 = \frac{5e}{0.5} = 10e$$

So: $\qquad\qquad 10e = 0.1g$

$$e = 0.01 \times 9.8 = 0.098$$

The particle is pulled downwards a further 0.2 m and released. The motion, while the string is taut, will be simple harmonic with amplitude 0.2 m.

To use the standard S.H.M. formulae, ω must be known. Consider P when it is a distance x below its equilibrium position. The acceleration will be $\ddot{x}$ downwards as shown.

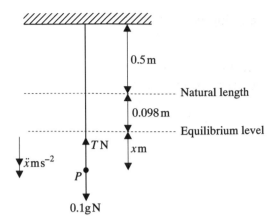

Using Newton's second law, $F = ma$, gives:

$$0.1g - T = 0.1\ddot{x} \qquad\qquad (1)$$

Using Hooke's law:

$$T = \frac{\lambda \times \text{extension}}{\text{natural length}}$$

with $\lambda = 5\,\text{N}$, extension $= (0.098 + x)\,\text{m}$, natural length $0.5\,\text{m}$ gives:

$$T = \frac{5(0.098 + x)}{0.5}$$

$$T = 10(0.098 + x)$$

And substituting this expression for T in equation (1) gives:

$$0.1 \times 9.8 - 10(0.098 + x) = 0.1\ddot{x}$$

$$0.98 - 0.98 - 10x = 0.1\ddot{x}$$

$$\ddot{x} = -\frac{10x}{0.1}$$

$$\ddot{x} = -100x$$

Comparing this with the standard S.H.M. equation $\ddot{x} = -\omega^2 x$ shows that $\omega = 10$.

To find the time that elapses before the string becomes slack use the equation

$$x = a\sin(\omega t + \alpha)$$

Substituting $x = a$ and $t = 0$ gives:

$$a = a\sin\alpha$$

So: $\sin \alpha = 1$

and: $\alpha = \dfrac{\pi}{2}$

Therefore: $x = a \sin\left(\omega t + \dfrac{\pi}{2}\right)$

or: $x = a \cos \omega t$

From above, $a = 0.2\,\text{m}$ and $\omega = 10$, so:

$$x = 0.2 \cos 10t$$

The string will become slack when there is no extension, that is when $x = -0.098\,\text{m}$

So: $-0.098 = 0.2 \cos 10t$

$$-0.49 = \cos 10t \qquad (2)$$

When the string *first* becomes slack $\frac{\pi}{2} < 10t < \pi$

So the required solution for equation (2) is

$$10t = 2.082$$
$$t = 0.2082$$

The string will become slack after $0.208\,\text{s}$.

While the string is slack the particle moves vertically under gravity alone. To calculate the time for which this happens you must first calculate the speed of the particle at the moment the string becomes slack.

Using $v^2 = \omega^2(a^2 - x^2)$ with $x = -0.098\,\text{m}$, $\omega = 10$ and $a = 0.2\,\text{m}$ gives:

$$v^2 = 10^2(0.2^2 - (-0.098)^2)$$
$$v = 1.743$$

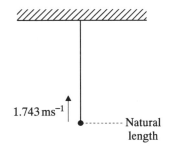

1.743 ms^{-1} ---- Natural length

For the motion under gravity the known quantities are $u = 1.743\,\text{m s}^{-1}$ and $a = -9.8\,\text{m s}^{-2}$. The string will become taut once more when $s = 0\,\text{m}$.

Using: $s = ut + \frac{1}{2}at^2$

gives: $0 = 1.743t - \frac{1}{2} \times 9.8t^2$

$$0 = (1.743 - 4.9t)t$$

The solutions are: $t = 0$ (the moment the string becomes slack)

or: $t = \dfrac{1.743}{4.9}$

$$t = 0.3557 \text{ (the moment the string}$$
$$\text{becomes taut once more)}$$

The further time that elapses before the string becomes taut once more is 0.356 s.

Exercise 2G

Whenever a numerical value of g is required, take $g = 9.8 \text{ m s}^{-2}$.

1 A particle of mass 2 kg is attached to the free end of an elastic spring of natural length 2 m and modulus 80 N which is hanging vertically. Determine the extension of the spring when the particle is hanging in equilibrium. The particle is pulled downwards a further 0.5 m and released. Calculate the period of the resulting oscillations and the maximum speed of the particle.

2 A particle of mass 0.5 kg is hanging in equilibrium attached to the free end of an elastic spring of natural length 0.8 m and modulus 60 N which is hanging vertically. The particle is pulled downwards and released. Calculate the period of the resulting oscillations. Given that the particle passes through its equilibrium position with speed 2 m s^{-1}, calculate the distance below the equilibrium position of the point from which the particle was released.

3 A particle of mass 1.5 kg is hanging in equilibrium attached to the free end of an elastic spring of natural length 1.2 m which is hanging vertically. The particle receives an impulse of 6 N s parallel to the spring. Show that the subsequent motion is simple harmonic.
Given that the period of the motion is $\dfrac{2\pi}{5}$ calculate the modulus of the spring and the amplitude of the motion.

4 A light elastic spring has one end A fixed and hangs vertically with a particle P of mass 0.2 kg attached to its free end. The particle is pulled downwards from its equilibrium position and released. P subsequently moves with simple harmonic motion of period $\dfrac{\pi}{3}$ s and the maximum and minimum distances of P below A are 1.5 m and 1 m. Calculate (a) the distance below A of the

centre of the oscillation (b) the amplitude of the oscillation
(c) the maximum speed of *P*.

5 A child's toy consists of a 'man' attached to a spring which is
fixed to a board. The mass of the man is 0.25 kg. When in
equilibrium with the spring vertical as shown in the diagram,
the spring is compressed by 0.02 m. The man is pushed
downwards a further 0.02 m and then released from rest.
Choose suitable models for the man and spring, stating clearly
the assumptions you make, and hence calculate the amplitude
and period of the resulting oscillations.

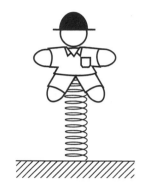

6 A piston of mass 2.6 kg moves in a smooth cylinder which is
fixed with its axis vertical. It is attached to the base of the
cylinder by a spring of natural length 15 cm and modulus 400 N.
Assuming that the spring is attached to the centres of the bases
of the piston and cylinder, choose suitable models for the piston
and spring stating clearly any further assumptions you make.
The piston is released from rest at a point where the spring has
length 12 cm. Calculate the period of the resulting oscillations
and the maximum speed of the piston.
State briefly how the model could be refined to enable
calculation of more accurate solutions (do not perform any
calculations).

7 A particle *P* of mass 0.3 kg is hanging in equilibrium attached to
one end of a light elastic string of natural length 0.6 m and
modulus 9N. *P* is pulled downwards a further 0.4 m and
released. Calculate the time that elapses before the string
becomes slack and the further time before the string is once
more taut.

8 A particle *P* of mass 2 kg is hanging in equilibrium attached to
one end of a light elastic string of natural length 1.2 m and
modulus 48 N. *P* is pulled downwards a distance 0.5 m and
released. When *P* has risen a distance 0.3 m the string is cut.
Calculate the greatest height *P* reaches above its equilibrium
position and the time taken to reach that greatest height.

9 A rock climber of mass 80 kg is attached to one end of a flexible
rope of natural length 20 m and modulus 45 000 N. She slips

from an overhang which is 10 m vertically below the point to which the other end of the rope is fixed.

(a) How could you model the climber and the rope to investigate her subsequent motion? State clearly any assumptions you would make about the rope and climber.

Assuming that the climber falls vertically without hitting the rock face, calculate (b) the speed of the climber when the rope becomes taut (c) the maximum distance of the climber below the point where she slipped (d) the time from slipping to reaching her lowest point.

SUMMARY OF KEY POINTS

1 For a particle of mass m moving in a straight line under the influence of a force $F = F(t)$

$$m\frac{dv}{dt} = F(t)$$

2 For a particle of mass m moving in a straight line under the influence of a force $F = G(x)$

$$m\frac{d}{dx}\left(\frac{1}{2}v^2\right) = G(x)$$

3 The impulse of a variable force $F(t)$ acting over the time interval t_1 to t_2 is:

$$\int_{t_1}^{t_2} F(t)\,dt$$

4 The work done by a variable force $G(x)$ which moves its point of application from x_1 to x_2 is:

$$\int_{x_1}^{x_2} G(x)dx$$

5 Newton's law of gravitation can be expressed by the equation:

$$F = \frac{GM_1M_2}{d^2}$$

where F is the force of attraction between two bodies of masses M_1 and M_2 which are a distance d apart.
G is the constant of gravitation.

6 Elastic strings and springs have a tension or thrust, T, given by:

$$T = \frac{\lambda x}{l}$$

where λ is the modulus of elasticity, x the extension or compression and l the natural length of the string or spring. This is known as Hooke's law.

7 A particle which moves on a straight line so that its acceleration is always towards a fixed point O in the line and is proportional to its displacement from O is said to move with simple harmonic motion or S.H.M.

This is written as: $\ddot{x} = -\omega^2 x$

The maximum displacement of the particle from O is the amplitude, a, of the motion.

The period of the motion is $\frac{2\pi}{\omega}$

The speed, v, at any point is given by

$$v^2 = \omega^2(a^2 - x^2)$$

and $x = a \sin \omega t$ if $x = 0$ when $t = 0$

$x = a \cos \omega t$ if $x = a$ when $t = 0$

$x = a \sin(\omega t + \alpha)$ if x is at some other point when $t = 0$

8 The period T of a simple pendulum is given by

$$T = 2\pi \sqrt{\left(\frac{l}{g}\right)}$$

where l is the length of the pendulum.

9 A particle attached to the end of an elastic spring which is displaced from its equilibrium position and then allowed to move under the influence of the tension or thrust only will move with S.H.M.

If the particle is attached to a string, the motion will be simple harmonic only while the string is taut.

Uniform circular motion

3

When a particle P is moving on a circular path with centre O you can specify its position by measuring the angle between the radius OP and a fixed radius OA.

In the study of circular motion angles are measured in radians. Radians are discussed in Book P1, chapter 7.

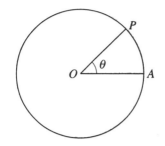

3.1 Angular speed

As P moves the angle θ changes. The rate at which θ changes with time is the **angular speed** ω of the radius OP. Since the angle θ is measured in radians the usual units for angular speed are radians per second (rad s^{-1}).

When it is necessary to distinguish between clockwise and anti-clockwise rotation the usual convention for angle measurement is used – that is, anticlockwise rotation is positive and clockwise rotation is negative.

The angular speed of the radius OP may also be described by the number of complete revolutions made in a given time interval. If the time interval is one minute the angular speed is measured in revolutions per minute (rev min^{-1}).

As ω is the rate of change of θ with respect to time it follows that:

$$\omega = \frac{\mathrm{d}\theta}{\mathrm{d}t}$$

When the angular speed ω is *constant* the particle is said to be moving in **uniform circular motion**.

It is important to remember that the angular speed ω refers to the speed at which the radius OP is turning. However, it is quite common to refer to the angular speed ω of the particle P about O even though strictly speaking this is not correct.

Example 1

A particle P is moving on a circular path centre O so that OP has a constant angular speed of $5 \, \text{rad s}^{-1}$. Find the angular speed of OP in rev min^{-1}.

As:
$$2\pi \, \text{rad} = 1 \, \text{rev}$$

So:
$$5 \, \text{rad s}^{-1} = \frac{5}{2\pi} \, \text{rev s}^{-1}$$

And:
$$1 \, \text{minute} = 60 \, \text{seconds}$$

So:
$$\frac{5}{2\pi} \, \text{rev s}^{-1} = \frac{5}{2\pi} \times 60 \, \text{rev min}^{-1}$$

$$= \frac{150}{\pi} \, \text{rev min}^{-1}$$

The angular speed of OP is $\dfrac{150}{\pi} \, \text{rev min}^{-1}$.

Example 2

A particle P is moving on a circular path centre O and is seen to make 10 complete revolutions in 4 minutes. Find the angular speed of P about O.

10 revolutions in 4 minutes is equivalent to $2.5 \, \text{rev min}^{-1}$. As above, $1 \, \text{rev} = 2\pi \, \text{rad}$ and $1 \, \text{min} = 60 \, \text{s}$

So:
$$2.5 \, \text{rev min}^{-1} = \frac{2.5 \times 2\pi}{60} \, \text{rad s}^{-1}$$

$$= \frac{\pi}{12} \, \text{rad s}^{-1}$$

The angular speed of P about O is $\dfrac{\pi}{12} \, \text{rad s}^{-1}$.

Angular and linear speed

Consider a particle P moving on a circular path centre O of radius r such that OP has constant angular speed ω. Take $\mathbf{i}$ and $\mathbf{j}$ to be unit vectors along and perpendicular to OA, the fixed direction from which θ is measured, as shown:

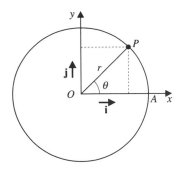

Suppose OP makes an angle θ with OA at time t seconds.

Since: $\qquad \dfrac{\mathrm{d}\theta}{\mathrm{d}t} = \omega \qquad$ and ω is constant,

integrating with respect to t gives:

$$\theta = \int \omega \, \mathrm{d}t$$
$$= \omega t + c$$

where c is an arbitrary constant.

Taking $\theta = 0$ when $t = 0$ gives $c = 0$ and hence:

$$\theta = \omega t$$

That is: $\qquad\qquad\qquad \angle AOP = \omega t$

Hence when OP has constant angular speed ω it turns through an angle ωt in time t seconds. Writing the position vector $\mathbf{r}$ of P relative to O in component form gives

$$\mathbf{r} = r \cos \omega t \mathbf{i} + r \sin \omega t \mathbf{j}$$

as $\theta = \omega t$.

In Book M1, chapter 2 it was shown how vectors written in component form can be differentiated. In Book P2, chapter 8 it was shown that:

$$\frac{\mathrm{d}}{\mathrm{d}t} (\cos \omega t) = -\omega \sin \omega t$$

and: $\qquad\qquad \dfrac{\mathrm{d}}{\mathrm{d}t} (\sin \omega t) = \omega \cos \omega t$

So differentiating $\mathbf{r}$ with respect to time gives:

$$\mathbf{v} = \dot{\mathbf{r}} = -r\omega \sin \omega t \, \mathbf{i} + r\omega \cos \omega t \, \mathbf{j}$$

The linear speed v of the particle is the modulus of $\mathbf{v}$.

Thus: $\qquad\qquad v = |\mathbf{v}| = |-r\omega \sin \omega t \, \mathbf{i} + r\omega \cos \omega t \, \mathbf{j}|$
$$= r\omega |-\sin \omega t \, \mathbf{i} + \cos \omega t \, \mathbf{j}|$$

$$= r\omega\sqrt{\{\sin^2 \omega t + \cos^2 \omega t\}}$$

But $\sin^2 \omega t + \cos^2 \omega t = 1$ as shown in Book P2, chapter 6 so that v is constant and

■ $$v = r\omega$$

When the radius is measured in metres and the angular speed in radians per second, the linear speed is in metres per second.

Direction of the linear velocity

If the particle P moving on a circle of radius r and centre O has position vector $\mathbf{r}$ relative to O given by

$$\mathbf{r} = r\cos \omega t\, \mathbf{i} + r\sin \omega t\, \mathbf{j}$$

then the velocity vector $\mathbf{v}$ of P is given by

$$\mathbf{v} = -r\omega \sin \omega t\, \mathbf{i} + r\omega \cos \omega t\, \mathbf{j}$$

as shown above.

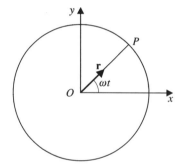

The gradient of the line OP is $\dfrac{r \sin \omega t}{r \cos \omega t} = \dfrac{\sin \omega t}{\cos \omega t}$

From the equation for $\mathbf{v}$ the gradient of the line representing the velocity vector $\mathbf{v}$ is

$$-\frac{r\omega \cos \omega t}{r\omega \sin \omega t} = -\frac{\cos \omega t}{\sin \omega t}$$

In Book P2, chapter 10 it was shown that:

if two lines have gradients m_1 and m_2 and

$$m_1 \times m_2 = -1$$

then the lines are perpendicular.

As : $$\frac{\sin \omega t}{\cos \omega t} \times \left(-\frac{\cos \omega t}{\sin \omega t}\right) = -1$$

it follows that the vectors $\overrightarrow{OP}$ and $\mathbf{v}$ are perpendicular. That is, the position vector $\mathbf{r}$ and the velocity vector $\mathbf{v}$ of P are perpendicular. Hence **the velocity is directed along the tangent to the circular path of P** as shown in the diagram below.

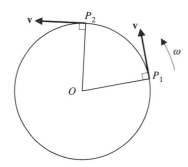

Angular speed is shown on diagrams with a curved arrow.

Example 3

A particle is moving on a circular path of radius 0.2 m, centre O, with constant angular speed $30 \, \text{rad s}^{-1}$ about O. Find its linear speed.

Using:
$$v = r\omega$$

with $r = 0.2$ and $\omega = 30 \, \text{rad s}^{-1}$ gives:

$$v = 0.2 \times 30$$

$$= 6$$

The linear speed of the particle is $6 \, \text{m s}^{-1}$.

Example 4

A particle is moving on a circular path of radius 0.75 m, centre O, with constant speed $5 \, \text{m s}^{-1}$. Find its angular speed about O in rev min^{-1}.

Using:
$$v = r\omega$$

with $r = 0.75 \, \text{m}$ and $v = 5 \, \text{m s}^{-1}$ gives:

$$5 = 0.75\omega$$

$$\omega = \frac{5}{0.75}$$

$$\omega = \frac{20}{3}$$

The particle's angular speed about O is $\dfrac{20}{3}$ rad s^{-1}.

So: angular speed $= \dfrac{20}{3} \times \dfrac{60}{2\pi} = \dfrac{200}{\pi}$ rev min^{-1}.

The particle's angular speed about O is $\dfrac{200}{\pi}$ rev min^{-1}.

Example 5

A particle is moving on a circular path of radius 0.5 m at a constant speed of 10 m s^{-1}. Find the time taken by P to complete 20 revolutions.

Using: $$v = r\omega$$

with $v = 10$ m s^{-1} and $r = 0.5$ m gives:

$$10 = 0.5\omega$$

$$\omega = 20$$

The angular speed is 20 rad s^{-1}.

Since: 20 revolutions $= 20 \times 2\pi = 40\pi$ radians

Using: $$\theta = \omega t$$

with $\theta = 40\pi$ rad and $\omega = 20$ rad s^{-1} gives:

$$40\pi = 20t$$

$$t = 2\pi = 6.28$$

The time taken is 6.28 seconds.

Exercise 3A

1 A particle is moving on a circular path with a constant angular speed of 3 rad s^{-1}. Find its angular speed in rev min^{-1}.

2 A particle is moving on a circular path with a constant angular speed of 4 rev min^{-1}. Find its angular speed in rad s^{-1}.

3 A particle moving on a circular path with constant speed is seen to complete 20 revolutions in 2 minutes. Find its constant angular speed (a) in rev min^{-1} (b) in rad s^{-1}.

4 A particle moving on a circular path with constant speed completes 5 revolutions in 40 seconds. Find its constant angular speed (a) in rev min^{-1} (b) in rad s^{-1}.

5 A particle is moving on a circular path of radius 0.5 m with constant angular speed 25 rad s^{-1}. Find its constant linear speed.

6   A particle is moving on a circular path of diameter 0.5 m with constant angular speed 3 rev min^{-1}. Find its constant linear speed.

7   The blade of a circular saw of radius 10 cm is rotating at 4000 rev min^{-1}. Find the speed of a point on its rim in m s^{-1}.

8   A particle is moving on a circular path of radius 0.3 m with a speed of 9 m s^{-1}. Find its angular speed in rad s^{-1} and the time it takes to complete 6 revolutions.

9   A river which is flowing at 3.6 m s^{-1} is turning a water wheel of radius 2 m. Assuming that the points on the rim of the wheel are moving at the speed of the water, calculate the angular speed of the wheel.

10 A car takes a bend by travelling along an arc of a circle of radius 35 m at 14 m s^{-1}. Calculate the constant angular speed of the car.

11 Anne and John are standing on a moving roundabout 75 cm and 125 cm respectively from the centre. John's speed is 2 m s^{-1}. Find (a) the angular speed of the roundabout, assumed constant, (b) Anne's speed.

3.2 Acceleration in circular motion

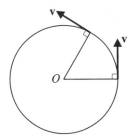

When a particle is moving on a circular path centre O at a constant speed its direction of motion is continually changing. So the velocity of the particle is changing and the particle therefore has an acceleration.

If a particle P is moving on a circle centre O of radius r, at time t seconds it has position vector $\mathbf{r}$ relative to O given by

$$\mathbf{r} = r \cos \omega t\, \mathbf{i} + r \sin \omega t\, \mathbf{j}$$

as shown in section 3.1.

Differentiating with respect to t gives:

$$\mathbf{v} = \dot{\mathbf{r}} = -r\omega \sin \omega t\, \mathbf{i} + r\omega \cos \omega t\, \mathbf{j}$$

Differentiating again with respect to t gives:

$$\mathbf{a} = \dot{\mathbf{v}} = \ddot{\mathbf{r}} = -r\omega^2 \cos \omega t\, \mathbf{i} - r\omega^2 \sin \omega t\, \mathbf{j}$$
$$\mathbf{a} = -\omega^2 (r \cos \omega t\, \mathbf{i} + r \sin \omega t\, \mathbf{j})$$

But:
$$\mathbf{r} = r \cos \omega t\, \mathbf{i} + r \sin \omega t\, \mathbf{j}$$

So: $$\mathbf{a} = -\omega^2\mathbf{r}$$

and the magnitude of $\mathbf{a}$, $|\mathbf{a}| = \omega^2 \times$ magnitude of $\mathbf{r}$

$$a = r\omega^2$$

Because of the minus sign, the direction of $\mathbf{a}$ is opposite to the direction of $\mathbf{r}$.

That is, the acceleration of P has magnitude $r\omega^2$ and is directed towards the centre of the circle.

As: $$v = r\omega$$

It follows that: $$\omega = \frac{v}{r}$$

and substituting for ω in $a = r\omega^2$ gives:

$$a = \frac{v^2}{r}$$

So the acceleration of a particle moving on a circular path centre O of radius r is given by:

■ $$a = r\omega^2 \quad \text{or} \quad a = \frac{v^2}{r}$$

and is directed towards the centre O of the circle.

Example 6
A particle is moving on a horizontal circular path of radius $0.5\,\text{m}$ with a constant angular speed of $4\,\text{rad s}^{-1}$. Calculate its acceleration.

Using: $$a = r\omega^2$$

with $r = 0.5\,\text{m}$ and $\omega = 4\,\text{rad s}^{-1}$ gives:

$$a = 0.5 \times 4^2$$

$$a = 8$$

The acceleration of the particle is $8\,\text{m s}^{-2}$ towards the centre of the circle.

Exercise 3B

1 A particle is moving on a horizontal circular path of radius $0.75\,\text{m}$ with a constant angular speed of $5\,\text{rev min}^{-1}$. Calculate the acceleration in m s^{-2} of the particle.

2   A particle is moving on a horizontal circular path of radius 2 m with a constant speed of $20 \, \text{m s}^{-1}$. Determine the acceleration of the particle.

3   A particle is moving on a horizontal circular path of radius 1.5 m. The acceleration of the particle is $150 \, \text{m s}^{-2}$ towards the centre of the circle. Calculate (a) the angular speed, (b) the linear speed of the particle.

4   A car takes a bend by travelling along an arc of a circle of radius 35 m at $14 \, \text{m s}^{-1}$. Calculate the acceleration of the car.

5   An insect flying around a light in a horizontal circle of radius 0.15 m is seen to make 30 revolutions per minute. Determine its acceleration.

6   Assuming the earth to be a sphere of radius 6370 km which makes one complete revolution in 24 hours, calculate the acceleration in m s^{-2} of a point on the equator.

3.3 Uniform motion of a particle moving in a horizontal circle

A particle which moves in a circle with constant speed has an acceleration directed towards the centre of the circle. There must be a force acting on the particle to produce this acceleration.

Using the equation of motion $F = ma$, along the radius, for a particle of mass m moving in a circle of radius r with constant speed v gives:

$$F = m \, \frac{v^2}{r} \quad \text{as} \quad a = \frac{v^2}{r}$$

So there is a force of magnitude $m \dfrac{v^2}{r}$ directed towards the centre of the circle. This force can arise in several ways.

(1) A particle which is attached to one end of a string whose other end is fixed and moves in a horizontal circle will experience a force due to the tension in the string.

(2) A bead which is threaded on a smooth circular wire and is made to move round the wire will experience a force due to the reaction of the wire. This will act along the inward radius, thus providing the required force towards the centre.

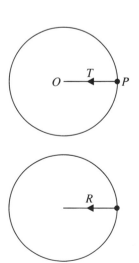

(3) A car travelling round a bend, which is an arc of a circle, on a horizontal road will experience a frictional force between the road and its tyres.

This force will be directed towards the centre O of the circular path and will produce the required acceleration if it is sufficiently large. However, there is a maximum frictional force that can result from the contact of two surfaces, as seen in Book M1, chapter 4. If this maximum frictional force is not sufficient to produce the required acceleration, the car will slip as it attempts to round the bend.

The force directed towards the centre of the circle is often referred to as the **centripetal force**.

Example 7

A particle P of mass 0.5 kg is attached to one end of a light inextensible string of length 0.25 m. The other end of the string is fixed to a point O of a smooth horizontal table. P moves with speed $2\,\mathrm{m\,s^{-1}}$ in a horizontal circle on the table. Calculate the tension in the string.

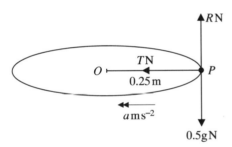

Let the tension in the string be T N and the acceleration towards the centre be $a\,\mathrm{m\,s^{-2}}$.

Using:
$$a = \frac{v^2}{r}$$

with $v = 2\,\mathrm{m\,s^{-1}}$ and $r = 0.25\,\mathrm{m}$ gives:

$$a = \frac{4}{0.25} = 16$$

The force required for the acceleration towards the centre is provided by the tension in the string.

Using:
$$F = ma$$

with $m = 0.5\,\mathrm{kg}$ and $a = 16\,\mathrm{m\,s^{-2}}$ gives:

$$T = 0.5 \times 16 = 8$$

The tension in the string is 8 N.

Example 8

A particle P of mass $0.5\,\text{kg}$ rests on a rough horizontal disc at a distance $0.1\,\text{m}$ from the centre. The disc is rotating about its centre. The coefficient of friction between the particle and the disc is $\frac{1}{4}$. Given that the particle is on the point of slipping, find the angular speed of the disc.

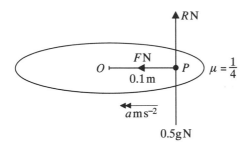

Here, the centripetal force is provided by the frictional force $F\text{N}$ between the particle and the disc. Let the (angular) speed of the disc be $\omega\,\text{rad s}^{-1}$ and the acceleration towards the centre be $a\,\text{m s}^{-2}$.

Using:
$$a = r\omega^2$$

with $r = 0.1\,\text{m}$ gives:
$$a = 0.1\omega^2$$

Using:
$$F = ma$$

with $m = 0.5\,\text{kg}$ and $a = 0.1\omega^2\,\text{m s}^{-2}$ gives:
$$F = 0.5 \times 0.1\omega^2$$
$$F = 0.05\omega^2$$

So the frictional force when the angular speed is $\omega\,\text{rad s}^{-1}$ is $0.05\omega^2\,\text{N}$.

Since the particle is on the point of slipping:
$$F = \mu R$$

(from Book M1 chapter 4) and resolving vertically (no motion) gives:
$$R = 0.5\text{g}$$

As $\mu = \frac{1}{4}$ it follows that:
$$0.05\omega^2 = \frac{1}{4} \times 0.5\,\text{g}$$
$$\omega^2 = \frac{1}{4} \times \frac{0.5}{0.05}\,\text{g}$$

We require the speed, that is the magnitude of ω, and so we take the positive square root. (The sign of ω indicates the direction of its velocity.)

Hence: $$\omega = 4.949$$

The angular speed of the disc when the particle is on the point of slipping is $4.95\,\text{m s}^{-1}$.

Example 9

A particle P of mass $0.25\,\text{kg}$ is attached to one end of a light elastic string of modulus $5\,\text{N}$ and natural length $0.75\,\text{m}$. The other end of the string is fixed to a point O on a smooth horizontal table. P moves in a horizontal circle of radius $1\,\text{m}$. Calculate the constant linear speed of P.

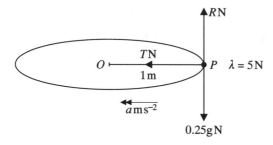

Let the speed of P be $v\,\text{m s}^{-1}$ and the tension in the string be $T\,\text{N}$. The tension can be obtained by using Hooke's law, which is explained in section 2.3.

Using Hooke's law: $$T = \frac{\lambda x}{l}$$

with $\lambda = 5\,\text{N}$, $l = 0.75\,\text{m}$ and $x = (1 - 0.75)\,\text{m} = 0.25\,\text{m}$ gives:

$$T = \frac{5 \times 0.25}{0.75}$$

$$T = \tfrac{5}{3}$$

The tension in the string is $\tfrac{5}{3}$ N.

Using: $$F = ma$$

with $m = 0.25\,\text{kg}$ and $F = T\,\text{N} = \tfrac{5}{3}\,\text{N}$ gives:

$$\frac{5}{3} = 0.25a$$

$$a = \frac{5}{3 \times 0.25}$$

$$a = \frac{20}{3}$$

The acceleration of the particle is $\frac{20}{3}\,\text{m s}^{-2}$.

Using:
$$a = \frac{v^2}{r}$$

with $a = \frac{20}{3}\,\text{m s}^{-2}$ and $r = 1\,\text{m}$ gives:

$$\frac{20}{3} = \frac{v^2}{1}$$

$$v^2 = \frac{20}{3}$$

and as $v \geqslant 0$:
$$v = 2.581$$

The speed of P is $2.58\,\text{m s}^{-1}$.

The conical pendulum

A particle may move in a horizontal circle attached to a string whose other end is fixed not at the centre O of the circle but at a point A vertically above the centre. This situation is called a 'conical pendulum' since the string describes the surface of a cone as the particle moves around the circle.

Example 10

A particle of mass 2 kg is attached to one end B of a light inelastic string AB of length 0.5 m. The other end A of the string is fixed. The particle moves with constant angular speed in a horizontal circle of radius 0.3 m whose centre O is vertically below A. Calculate the tension in the string and the angular speed of the particle.

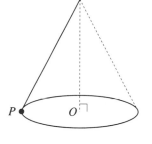

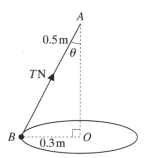

Let the tension in the string be T N and the angular velocity be ω rad s^{-1}.

The motion is taking place in a horizontal plane and the particle is not moving vertically.

So, resolving vertically:

$$T\cos\theta = 2g$$

From the diagram you can see that triangle OAB is a (3,4,5) triangle

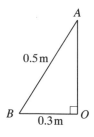

so: $$OA = 0.4\,\text{m}$$

and: $$\cos\theta = \tfrac{4}{5}$$

Hence: $$T \times \tfrac{4}{5} = 2g$$

$$T = \frac{10}{4}\,g = 24.5$$

The tension in the string is 24.5 N.

The particle is moving in a horizontal circle. Using the equation of motion:

$$F = ma \qquad \text{along } BO$$

gives: $$T\sin\theta = 2r\omega^2$$

Again from triangle OAB:

$$\sin\theta = \frac{3}{5}$$

and as $T = 24.5\,\text{N}$ and $r = 0.3\,\text{m}$ this gives:

$$24.5 \times \tfrac{3}{5} = 2 \times 0.3\omega^2$$

$$\omega^2 = \frac{24.5 \times 3}{5 \times 2 \times 0.3}$$

$$\omega = 4.949$$

as speed must be positive.

The angular speed is $4.95\,\text{rad s}^{-1}$.

Example 11

A ring R of mass 0.1 kg is threaded on a light inelastic string AB of length 3.5 m. The ends A and B of the string are attached to two fixed points 2.5 m apart in a vertical line. The ring is moving in a horizontal circle with constant angular speed and $\angle ARB$ is 90°.

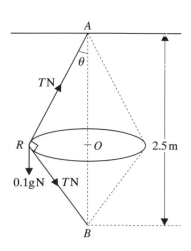

(a) Suggest a suitable model for the ring so that the tension in the string can be calculated.
(b) By considering the forces acting on the ring show that

$$\frac{AR}{RB} > 1$$

(c) Prove that $AR = 2\,\text{m}$ and find the radius of the circular path described by the ring.
(d) Calculate the tension in the string and the angular speed of the ring.

As the ring is only *threaded* on the string and there is no knot at the ring the string is continuous. Consequently the tensions in AR and BR are the same. Let these tensions be T N and let $\angle RAB$ be θ.

(a) The ring can be modelled as a particle as it is small compared with the length of the string.

(b) As there is no vertical motion, resolving vertically gives:

$$T\cos\theta = 0.1\text{g} + T\sin\theta \qquad (1)$$

So:

$$T(\cos\theta - \sin\theta) = 0.1\text{g}$$

As $T > 0$ it follows that:

$$\cos\theta - \sin\theta > 0$$

and so:

$$\cos\theta > \sin\theta$$

As:

$$\frac{\sin\theta}{\cos\theta} = \tan\theta$$

(Book P1, chapter 7) it follows that:

$$\tan\theta < 1$$

From triangle ARB, which is right-angled at R:

$$\tan\theta = \frac{BR}{AR}$$

And so:

$$\frac{BR}{AR} < 1$$

or:

$$\frac{AR}{BR} > 1$$

(c) Let AR be x m.

Then $BR = (3.5 - x)$ m as the string is 3.5 m long. As triangle ARB is right-angled, using Pythagoras' theorem gives:

$$AR^2 + BR^2 = AB^2$$
$$x^2 + (3.5 - x)^2 = 2.5^2$$
$$x^2 + 12.25 - 7x + x^2 = 6.25$$
$$2x^2 - 7x + 6 = 0$$
$$(2x - 3)(x - 2) = 0$$
$$x = \tfrac{3}{2} \quad \text{or} \quad x = 2$$

The two solutions for x are the lengths of AR and BR as their sum is 3.5.

From part (b): $\qquad\qquad AR > BR$

So: $\qquad\qquad\qquad AR = 2m$

To calculate the radius of the circle described by R consider triangle AOR.

Let the radius be r m.

Then: $\qquad\qquad\qquad r = AR \sin\theta$

From triangle ARB: $\qquad \sin\theta = \dfrac{RB}{AB} = \dfrac{1.5}{2.5} = \dfrac{3}{5}$ $\qquad$ (2)

So: $\qquad\qquad\qquad r = 2 \times \frac{3}{5} = \frac{6}{5}$

The radius of the circular path is $1\frac{1}{5}$ m.

(d) Equation (1) gives: $\quad T\cos\theta = 0.1g + T\sin\theta$

Equation (2) gives: $\qquad \sin\theta = \frac{3}{5}$

and hence: $\qquad\qquad \cos\theta = \frac{4}{5}$

So: $\qquad\qquad T \times \frac{4}{5} - T \times \frac{3}{5} = 0.1g$

$$\tfrac{1}{5}T = 0.1 \times 9.8$$

$$T = 4.9$$

The tension is 4.9 N.

To find the angular speed consider the equation of motion along the inward radius.

Using: $\qquad\qquad\qquad F = ma$

with $m = 0.1$ kg gives: $\qquad T\sin\theta + T\cos\theta = 0.1r\omega^2$

Substituting $T = 4.9$, $\sin\theta = \frac{3}{5}$, $\cos\theta = \frac{4}{5}$ and $r = \frac{6}{5}$ gives:

$$4.9 \times (\tfrac{3}{5} + \tfrac{4}{5}) = 0.1 \times \tfrac{6}{5} \times \omega^2$$

$$\omega^2 = \frac{4.9 \times 7}{0.1 \times 6}$$

$$\omega = 7.560 \quad \text{(as speed must be positive)}$$

The angular speed is 7.56 rad s^{-1}.

Example 12

A child's toy consists of a small sphere P of mass m joined to two rods AP, BP each of length l.

The other ends of the rods are a distance l apart on a third rod as shown in the diagram. AP and BP can rotate about AB. Assuming that the rods AP and BP are light and AB is vertical, calculate the tensions in the rods AP and BP when P is moving in a circle with

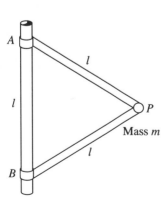

speed $\sqrt{(3gl)}$. State clearly the model you have used for P in your solution and any other assumptions made.

As P is a small sphere it can be modelled as a particle. It is also assumed that the rods AP and BP rotate freely about AB, that is there are no frictional forces at A and B.

Let the tensions in AP and BP be T_1 and T_2 respectively. Let P move in a circle centre O, the mid-point of AB. Triangle APB is equilateral as all sides have length l. Hence $\angle A = 60°$ and $OP = l\cos 30°$.

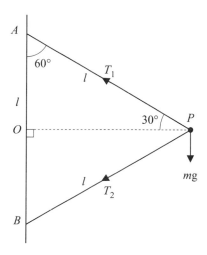

Using the equation of motion, $F = ma$, along the radius OP gives:

$$T_1 \cos 30° + T_2 \cos 30° = m\frac{v^2}{r}$$

But $r = l\cos 30°$ from above and $v = \sqrt{(3gl)}$

So:
$$T_1 \cos 30° + T_2 \cos 30° = m \times \frac{3gl}{l\cos 30°}$$

$$T_1 \cos^2 30° + T_2 \cos^2 30° = 3mg$$

But $\cos 30° = \dfrac{\sqrt{3}}{2}$ and so $\cos^2 30° = \frac{3}{4}$

Hence:
$$\frac{3T_1}{4} + \frac{3T_2}{4} = 3mg$$

and so:
$$T_1 + T_2 = 4mg \qquad (1)$$

There is no vertical motion, so resolving vertically gives:

$$T_1 \cos 60° = T_2 \cos 60° + mg$$
$$\tfrac{1}{2}T_1 = \tfrac{1}{2}T_2 + mg$$
$$T_1 - T_2 = 2mg \qquad (2)$$

Adding equations (1) and (2) gives:

$$2T_1 = 6\,mg$$
$$T_1 = 3\,mg$$

And subsituting this value of T_1 in either equation gives:

$$T_2 = mg$$

The tensions in AP and BP are $3\,mg$ and mg respectively.

Exercise 3C

Whenever a numerical value of g is required take $g = 9.8\,\mathrm{m\,s^{-2}}$.

1 A particle P of mass $1\,\mathrm{kg}$ is attached to one end of a light inextensible string of length $0.3\,\mathrm{m}$. The other end of the string is fixed to a point O of a smooth horizontal table. P moves with speed $5\,\mathrm{m\,s^{-1}}$ in a circle on the table. Calculate the tension in the string.

2 A particle P of mass $0.2\,\mathrm{kg}$ is attached to one end of a light inextensible string of length $0.5\,\mathrm{m}$. The other end of the string is fixed to a point O of a smooth horizontal table. P describes horizontal circles, centre O, on the table. Given that the tension in the string is $2.5\,\mathrm{N}$, calculate the speed of the particle.

3 A particle P of mass $0.2\,\mathrm{kg}$ rests on a rough horizontal disc at a distance of $0.3\,\mathrm{m}$ from the centre. The coefficient of friction between the particle and the disc is $\frac{1}{3}$. Given that the particle is on the point of slipping, calculate the angular speed of the disc.

4 A particle P of mass $0.5\,\mathrm{kg}$ rests on a rough horizontal disc at a distance of $0.25\,\mathrm{m}$ from the centre. When the disc is rotating with angular speed $5\,\mathrm{rad\,s^{-1}}$ P is on the point of slipping. Calculate the value of the coefficient of friction between P and the disc.

5 A smooth wooden hoop of radius $1\,\mathrm{m}$ rests on a smooth horizontal surface. A bead of mass $0.2\,\mathrm{kg}$ is moving in a horizontal circle in contact with the inner surface of the hoop. The bead maintains a constant speed of $6\,\mathrm{m\,s^{-1}}$. Calculate the magnitude of the reaction between the hoop and the bead.

6 A particle P of mass $2\,\mathrm{kg}$ is attached to one end of a light elastic string of modulus $20\,\mathrm{N}$ and natural length $0.5\,\mathrm{m}$. The other end

of the string is fixed to a point O on a smooth horizontal surface. P moves in a horizontal circle with centre O. Given that the speed of P is $10\,\mathrm{m\,s^{-1}}$, calculate the radius of the circle.

7 A particle of mass $0.2\,\mathrm{kg}$ is attached to the end C of two light inextensible strings AC and BC both of length $1\,\mathrm{m}$. The other ends A and B are attached to opposite ends of a diameter of a smooth disc of radius $0.8\,\mathrm{m}$ which is free to rotate in a horizontal plane about its centre. The disc rotates with angular speed $10\,\mathrm{rad\,s^{-1}}$ and P remains in contact with the disc.
(a) Calculate the tension in each string. (b) Given that the strings will break if the tension exceeds $30\,\mathrm{N}$, calculate the maximum speed at which the disc can rotate without the strings breaking.

8 A particle of mass m is attached to one end of a light elastic string of natural length l and modulus $5\,mg$. The other end of the string is fixed to a point O on a smooth horizontal table. The particle moves on the table in a circular path of radius $\dfrac{6l}{5}$ and centre O. Show that the constant speed u of the particle is given by

$$u = \sqrt{\frac{6gl}{5}}$$

9 OB is a light rigid rod of length $5l$. Particles of masses $3m$ and $4m$ are attached at points A and B of the rod, where OA is $2l$. The rod is made to rotate about O in a horizontal plane with constant angular speed ω. Calculate the ratio of the tensions in OA and AB.

10 A light rigid rod AB of length $6l$ has particles of masses $2m$ and $4m$ attached at A and B respectively. The rod rotates with constant angular speed ω in a horizontal plane about point D of the rod. The tensions in AD and DB are equal. Show that

$$AD = 4l$$

11 A bead P of mass $0.8\,\mathrm{kg}$ is attached to one end of a light inextensible string and rests on a smooth horizontal table. The string passes through a small smooth hole, O, in the table and carries a mass of $0.5\,\mathrm{kg}$ which hangs freely. P is moving on the

table in a circle of radius 0.4 m and centre O. Calculate the angular speed of P.

12 A bead B of mass 0.8 kg is threaded on a smooth circular wire centre O and radius 0.4 m which is fixed in a horizontal plane. A light inextensible string is attached to B and passes through a small smooth horizontal ring at O and carries a mass of 0.4 kg which is hanging freely. B is moving on the wire with a constant speed of $5\,\mathrm{m\,s^{-1}}$. Calculate the magnitude and direction of the reaction of the wire on B.

13 A glass marble of mass m is moving in a horizontal circle round the inside surface of a smooth hemispherical bowl of radius r. The centre of the circle is at a distance $\frac{1}{2}r$ below the centre of the bowl. State a suitable model for the marble and hence find the magnitude of the reaction between the marble and the bowl and the speed of the marble.

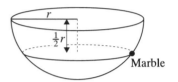

14 A boy sets his toy car running round the smooth sloping side of a circular sand-pit. The car moves in a horizontal circle of radius 0.75 m and the sides of the sand-pit are inclined at an angle arcsin $\frac{4}{5}$ to the horizontal. State suitable models for the car and the side of the sand-pit and hence calculate the speed of the car.

15 A particle of mass 0.8 kg is attached to end B of a light inelastic string AB of length 0.6 m, the other end of which is fixed at A. The particle moves with a constant angular speed in a horizontal circle of radius 0.2 m whose centre is vertically below A. Calculate the tension in the string and the angular speed of the particle.

16 A particle of mass 1.5 kg is attached to end B of a light inelastic string AB of length 0.5 m, the other end of which is fixed at A. The particle moves in a horizontal circle whose centre is vertically below A with a constant angular speed. The particle takes $\frac{\pi}{4}$ s to complete one revolution. Calculate the tension in the string and the radius of the circular path.

17 A particle P of mass 2 kg is attached to one end of a light inextensible string of length 1.2 m. The other end of the string is fixed at a point A which is 0.6 m above a smooth horizontal table. P moves on the table on a circular path whose centre O is

vertically below A. Given that the string is taut and the angular speed of P about A is $3\,\mathrm{rad\,s^{-1}}$ calculate (a) the tension in the string (b) the normal reaction between the particle and the table.

18 The end A of a light inextensible string of length l is attached to a fixed point. A particle P of mass m is attached to the free end. P moves with constant speed $2v$ in a horizontal circle with centre vertically below A and the string remains taut, inclined at an angle ϕ to the vertical. (a) Find v^2 in terms of l, g and ϕ. (b) Given also that the string breaks when the tension exceeds $4mg$, show that $v \leqslant \frac{1}{4}\sqrt{15lg}$.

19 A light elastic string AB has natural length l and modulus $\dfrac{20mg}{3}$.

The end A is fixed and a particle of mass m is attached to the end B.

The particle is made to move in a horizontal circle at a constant angular speed with the string inclined at an angle $\arcsin\frac{4}{5}$ to the vertical. Find, in terms of l and g, the angular speed of the particle.

20 A light inextensible string AB of length $3l$ has its ends fixed to two points A and B which are in a vertical line with A a distance l above B. A smooth ring of mass m is threaded on the string and is made to move in a horizontal circle centre B with a constant speed. (a) State a suitable model for the ring and hence calculate the tension in the string and the speed of the ring. (b) Explain briefly the significance of the statement 'The ring is threaded on the string' in the solution of the problem.

21 A small smooth ring of mass m is threaded on a light inelastic string of length $9l$, whose ends are fixed at a point O on a smooth horizontal table and at a point A which is at a height $6l$ vertically above O. The ring moves in a horizontal circle on the table with the string taut. Given that the ring is moving at a constant angular speed $\dfrac{1}{2}\sqrt{\left(\dfrac{g}{l}\right)}$ calculate the magnitude of the reaction between the ring and the table.

22 A particle P is attached to the ends of two identical light inelastic strings which have their other ends fixed to two points A and B where A is $2l$ above B in a vertical line. P moves in a

horizontal circle with constant angular speed $4\sqrt{\dfrac{g}{l}}$ and both strings are taut. Calculate the ratio of the tensions in AP and BP.

23 A particle P is attached to the ends of two elastic strings which have the same natural length l. The other ends of the strings are fixed to two points A and B where A is $3l$ above B in a vertical line. The modulus of the string BP is one third that of AP. P moves in a horizontal circle with centre at the mid-point of AB and both strings taut. Find the angular speed of P in terms of g and l.

SUMMARY OF KEY POINTS

1 When a particle P is moving on a circle centre O and the angle between OP and a fixed radius OA is θ, the angular speed of OP is ω where

$$\omega = \frac{\mathrm{d}\theta}{\mathrm{d}t}$$

2 Angular speed is usually measured in radians per second $(\mathrm{rad\,s^{-1}})$ or revolutions per minute $(\mathrm{rev\,min^{-1}})$.

3 The linear speed v of the particle is directed along the tangent to the circular path.

4 The linear and angular speeds are connected by the relationship

$$v = r\omega$$

where r is the radius of the circular path.

5 When r is measured in metres and ω in radians per second, v is in metres per second.

6 A particle moving in a circular path at a constant speed has an acceleration of magnitude a directed towards the centre of the circle where a is given by

$$a = r\omega^2 \text{ or } a = \frac{v^2}{r}$$

7 The resultant force $\mathbf{F}$ acting on a particle of mass m moving on a circular path with constant speed has magnitude $mr\omega^2$ or $\dfrac{mv^2}{r}$ and is directed towards the centre of the circle.

Review exercise

Whenever a numerical value of g or G is required take $g = 9.8\,\mathrm{m\,s^{-2}}$ and $G = 6.67 \times 10^{-11}\,\mathrm{kg^{-1}\,m^3\,s^{-2}}$.

1   A particle moves along Ox in the positive direction. At time $t = 0$, its speed is $8\,\mathrm{m\,s^{-1}}$. At time $t\,\mathrm{s}$, its acceleration is $3e^{2t}\,\mathrm{m\,s^{-2}}$ in the positive direction. Calculate, in metres to 3 significant figures, the distance the particle moves in the first 2 seconds.   [L]

2   A particle travels along a straight line which passes through A and B. During the motion the velocity of the particle is $6t + t^2$ metres per second, where t seconds is the time measured from a certain instant. The particle passes through A when $t = 2$ and through B when $t = 5$. Find, in terms of t, the acceleration of the particle and its distance from A at any instant during the motion. Calculate, also, the distance of B from A.   [L]

3   A particle starts with speed $20\,\mathrm{m\,s^{-1}}$ and moves in a straight line. The particle is subjected to a resistance which produces a retardation which is initially $5\,\mathrm{m\,s^{-2}}$ and which increases uniformly with the distance moved, having a value of $11\,\mathrm{m\,s^{-2}}$ when the particle has moved a distance of $12\,\mathrm{m}$. Given that the particle has speed $v\,\mathrm{m\,s^{-1}}$ when it has moved a distance of $x\,\mathrm{m}$, show that, while the particle is in motion,

$$v\frac{\mathrm{d}v}{\mathrm{d}x} = -(5 + \tfrac{1}{2}x)$$

Hence, or otherwise, calculate the distance moved by the particle in coming to rest.   [L]

4   A particle moving along a straight line starts at time $t = 0$ seconds with a velocity $4\,\mathrm{m\,s^{-1}}$. At any subsequent time t seconds the acceleration of the particle is $(6t - 8)\,\mathrm{m\,s^{-2}}$. Find

(a) the distance the particle moves before first coming to instantaneous rest,

(b) the total time T seconds taken by the particle to return to the starting point,

(c) the greatest speed of the particle for $0 \leqslant t \leqslant T$.   [L]

5 At time t seconds a particle moving on a straight line with velocity $v\,\text{m}\,\text{s}^{-1}$ experiences an acceleration $kt\,\text{m}\,\text{s}^{-2}$, where k is a positive constant. When $t = 0$, $v = 4$ and when $t = 10$, $v = 12$.

(a) Show that $k = \frac{4}{25}$.

(b) Find the distance covered by the particle in the interval $5 \leqslant t \leqslant 10$.   [L]

6 A particle moves on the positive x-axis. The particle is moving towards the origin O when it passes through the point A, where $x = 2a$, with speed $\sqrt{(k/a)}$, where k is constant. Given that the particle experiences an acceleration $k/(2x^2) + k/(4a^2)$ in a direction away from O, show that it comes instantaneously to rest at a point B, where $x = a$.

Immediately the particle reaches B the acceleration changes to $k/(2x^2) - k/(4a^2)$ in a direction away from O. Show that the particle next comes instantaneously to rest at A.   [L]

7 A particle moves in a straight line so that its distance, s metres, from the origin after time t seconds is given by $s = t + \sin 2t$.

Find expressions for

(a) the velocity $v\,\text{m}\,\text{s}^{-1}$

(b) the acceleration, $f\,\text{m}\,\text{s}^{-2}$

Hence show that

$$f^2 = 16 - 4(v - 1)^2.$$ [L]

8 A particle P starts from rest at a fixed point O and moves on a horizontal plane in a straight line Ox. At time $t\,\text{s}$ after leaving O the acceleration of P is $(15 + 4t - 3t^2)\,\text{m}\,\text{s}^{-2}$. The particle comes to instantaneous rest at the point A.

(a) Show that P takes $5\,\text{s}$ to reach A from O.

(b) Calculate the distance between O and A, giving your answer to the nearest m.

The mass of P is $0.2\,\text{kg}$. As P moves from O to A during the interval $0 \leqslant t \leqslant 5$, calculate

(c) the greatest kinetic energy of P, giving your answer in J,

(d) the greatest value of the magnitude of the force exerted on P, giving your answer in newtons.   [L]

9 A ball of mass $0.1\,\text{kg}$ falls vertically from rest in a tank of liquid. The ball has speed $v\,\text{m s}^{-1}$ at time t seconds and the resistance to motion is $0.98(1 - e^{t})\,\text{N}$.

(a) Show that the acceleration of B at time t seconds is:

$$9.8e^{t}\,\text{m s}^{-2}$$

(b) Find v in terms of t.

(c) State any assumptions you made about the ball when modelling its motion.

10 A particle P of mass m is dropped from a height h above the surface of the liquid in a container and falls freely under gravity until it enters the liquid. As P enters the liquid, $\frac{1}{4}$ of its kinetic energy is lost. At time t after entering the liquid P has fallen a vertical distance x and the resistance to motion is $3mx^{2}$. Calculate the speed of P when it has just entered the liquid and hence show that its speed at time t later is given by:

$$v^{2} = 2gx - 2x^{3} + \tfrac{3}{2}gh$$

11 A packet of mass $2\,\text{kg}$ is dropped from the top of a tower which is $60\,\text{m}$ high.

(a) Calculate the time taken by the packet to reach the ground, stating clearly any assumptions you make about the packet and the forces acting on it.

(b) The model can be refined by including the resistive force acting on the packet. Given that this resistive force has magnitude $19.6(1 - e^{\frac{-kt}{2}})\,\text{N}$ at time t seconds after the packet was dropped, find an expression for the speed $v\,\text{m s}^{-1}$ of the packet at time t seconds.

12 A particle P of mass $0.5\,\text{kg}$ moves away from the origin O along the positive x-axis under the action of a force directed away from O. At time t seconds after P leaves O the force has magnitude $(10 - 2t)\,\text{N}$. When $t = 0$ P has speed $5\,\text{m s}^{-1}$ and when $t = 3$ the force is removed. Calculate:

(a) the magnitude of the impulse given to the particle by the force between $t = 0$ and $t = 3$,

(b) the final speed of the particle.

13 A particle of mass 0.4 kg is initially at rest on a smooth horizontal surface. A horizontal force of magnitude $(2x + 1)$ N, where x m is the distance of the particle from its initial position, acts on the particle causing it to move in a straight line. Calculate:

(a) the work done by the force in moving the particle 3 m from its starting point,

(b) the speed of the particle when $x = 3$.

14 Assuming the earth to be a sphere of radius 6400 km calculate, in newtons, the magnitude of the gravitational force on an object of mass 80 kg which is 600 km above the surface of the earth. State clearly the models used for the earth and the object. Assuming there are no frictional forces acting, calculate the speed of the object if it is orbiting the earth at a constant height of 600 km above the surface.

15 A satellite is travelling in a circular orbit above the equator. Assuming the earth to be a sphere of radius 6370 km and given that the period of the satellite is 24 hours, find the height of the satellite above the earth's surface.

16 A spacecraft is orbiting the moon at a constant height of 80 km. Assuming the moon to be a sphere of radius 1740 km and mass 7.35×10^{22} kg, calculate the period of orbit of the spacecraft.

17 Assuming the earth to be a sphere of radius 6370 km, calculate the speed and height above the earth's surface of a satellite which takes 150 minutes to complete one circuit of the earth.

18 An elastic string of natural length l has modulus of elasticity λ. One end of the string is fixed and the string hangs in equilibrium with a particle of mass m attached to the lower end. Find the stretched length of the string. [L]

19 A spring whose natural length is 200 mm has a length of 220 mm when supporting in equilibrium a mass of 10 kg. Calculate its length when supporting in equilibrium a mass of 25 kg. [L]

20 A particle of mass M rests suspended from a peg by a light elastic string. When $M = 1.5$ kg the length of the string is 0.45 m. When $M = 2$ kg the length of the string is 0.5 m. Find the unstretched length of the string. [L]

21 State Hooke's Law for an elastic string.

An elastic string of natural length 1.2 m hangs vertically from a point O and has a body of mass 3 kg attached at the other end. The string hangs in equilibrium with the body 1.5 m below O. Calculate the modulus of elasticity of the string.

A new position of equilibrium, with the string inclined at an angle of 60° to the vertical, is maintained by a force applied to the body in a direction at right angles to the string.

Calculate

(a) the force applied,

(b) the tension in the string,

(c) the length of the string in this new position.   [L]

22 Two light elastic strings AB and CD, each with the same natural length c, have moduli of elasticity $9Mg$ and $12Mg$ respectively. The ends B and C are joined together, the end A is tied at a fixed point and a particle of mass M is attached to D. The particle hangs in equilibrium with the strings vertical. Find, in terms of c, the length of AD.   [L]

23 A particle P, of mass 6 kg, rests on a rough horizontal table at a distance of 15 cm from a vertical rod which is fixed to the table. A ring Q is free to move on the vertical rod and is attached to P by a light elastic string of natural length 15 cm. The string is such that when a mass of 3 kg is suspended freely at its lower end the extension is 15 cm. The ring Q is slowly raised up the vertical rod and has reached a height of 20 cm above the table when P is just about to slide on the table.

Find (a) the tension in the string.

(b) the normal reaction between P and the table.

Prove that the coefficient of friction between P and the table is $\frac{3}{11}$.   [L]

24

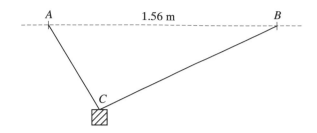

Two points A and B are 1.56 m apart in a horizontal line (see diagram). A body of mass 1.3 kg is suspended from A and B by two light elastic strings AC and BC. In the equilibrium position AC is perpendicular to BC and tan $ABC = 5/12$.

Calculate, by resolving horizontally or otherwise, the ratio of the tensions in the two strings, and find the magnitude of the tension in the string AC.

Given that the unstretched length of the string AC is 54 cm, calculate the value of the modulus of elasticity of this string.

[L]

25 A light elastic string, of natural length l and modulus of elasticity $4mg$, has one end tied to a fixed point A. The string passes over a fixed smooth peg B and at the other end a particle P, of mass m, is attached. The particle hangs in equilibrium. The distance between A and B is l and AB is inclined at $60°$ to the vertical as shown in the diagram.

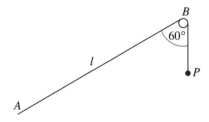

(a) Find, in terms of l, the length of the vertical portion BP of the string.

(b) Show that the magnitude of the force exerted by the string on the peg is $mg\sqrt{3}$.

[L]

26 A particle of mass m moves along the positive x-axis. It is acted upon by a force directed towards the origin and of magnitude mn^2x. Write down an equation governing the motion. Given that at time $t = 0$ the particle is at the origin and that the maximum distance of the particle from the origin is a, find x in terms of a, n and t.   [L]

27 A particle is moving in a straight line so that at time t it is a distance s from a fixed point where

$$s = a \cos nt \, ,$$

a and n being constants.

Find

(a) the velocity v and the acceleration f at time t,

(b) an expression for $v^2 - fs$, and show that it is constant,

(c) an expression for $\dfrac{df}{dt}$ and show that it is proportional to v.

Describe the motion of the particle along the straight line. [L]

28 A particle moves in a straight line with simple harmonic motion of period 4π seconds about a centre O. The speed of the particle is zero when it is at a distance $2\,\text{m}$ from O.

Find the speed when it is $1\,\text{m}$ from O.   [L]

29 A particle P starts from the point O at time $t = 0$ and moves in a straight line with velocity v, where

$$v = k\omega t \sin 2\omega t,$$

k and ω being constant.

Find

(a) the acceleration of P when $t = \pi/(2\omega)$,

(b) the distance OP when P comes momentarily to rest for the first time after leaving O.   [L]

30 A particle is moving along a straight line through a fixed point O and its displacement s metres, from O, at time t seconds, is given by the relation $s = 5 \cos 2t$. Obtain expressions for the velocity and the acceleration of the particle at time t seconds and show how this acceleration is related to the displacement.

When $t = \dfrac{\pi}{6}$ the particle is at the point A and is moving towards O. Find

(a) the distance OA,

(b) the time when the particle next reaches O,

(c) the velocity of the particle when it passes through O,

(d) the acceleration of the particle when it reaches its furthest point from A,

(e) the time, measured from $t = \dfrac{\pi}{6}$, that elapses before the particle next passes through A. [L]

31 A particle is moving along a straight line and its distance s metres from a fixed point O on the line at the instant t seconds after it was at O is $a \sin nt$ metres. Find the velocity and the acceleration of the particle t seconds after it was at O. The maximum speed of the particle is $0.6\,\mathrm{m\,s^{-1}}$ and its maximum acceleration is $2.4\,\mathrm{m\,s^{-2}}$. Find the values of a and n. Find also the time which elapses between the instant when the particle leaves O and when the particle is next at O. [L]

32 One end of a light elastic string, of natural length l, is attached to a fixed point O. To the other end of the string is attached a particle of mass m. When the particle hangs in equilibrium the length of the string is $3l/2$.

(a) The particle is pulled vertically downwards from its equilibrium position and released from rest when the length of the string is $7l/4$. Find the period and the amplitude of the subsequent oscillations.

(b) If instead the particle is released from rest at a point A vertically below O, where $OA = 9l/4$, show that the time taken for the string to first become slack is

$$\left(\frac{l}{2g}\right)^{\frac{1}{2}}\left[\pi - \arccos\left(\frac{2}{3}\right)\right].$$
[L]

33 (a) The level of the water in a harbour rises and falls vertically in simple harmonic motion. The depth of the water at low tide is $5.5\,\mathrm{m}$ and the depth at high tide is $11.5\,\mathrm{m}$. A boat can enter the harbour when the depth of the water is $7\,\mathrm{m}$. Low tide is at 0740 hours and high tide is at 1355 hours. Calculate the earliest time after 0740 hours at which the boat can just enter the harbour.
[L]

(b) Describe briefly the model you used for the boat when investigating the motion in part (a) of this question.

34 A girl sitting on a swing is pulled aside and released. State clearly the assumptions you are making if you consider her motion to be that of a simple pendulum. Given that these assumptions are valid and that the ropes of the swing are of length 2.25 m, calculate the period of oscillation.

35 A particle P of mass m hangs in equilibrium at one end of a vertical elastic string, the other end of which is attached to a fixed point. The extension of the string is e. If P is given a further extension, a, where $a < e$, and then released, find the period of its subsequent motion.   [L]

36 A light elastic spring, of modulus $4mg$ and natural length l, has one end attached to a fixed point and carries a scale pan, of mass m, at the other end. Show that the period T of small oscillations of the pan about its equilibrium position is

$$2\pi\sqrt{(l/4g)}.$$

When the pan is hanging in equilibrium, a particle of mass km is gently placed in the pan. Given that the periodic time of the ensuing motion is $2T$ find the value of k and the amplitude of the oscillation.   [L]

37 A light elastic string of natural length l has one end A attached to a fixed point. A particle of mass m is attached to the other end B. When the particle hangs in equilibrium AB has length $\dfrac{5l}{4}$. The particle is pulled down vertically a further distance a and released from rest. Given that $a \leqslant \dfrac{l}{4}$ show that B first comes to instantaneous rest after a time $\tfrac{1}{2}\pi\sqrt{\left(\dfrac{l}{g}\right)}$ and find its greatest speed during this motion.

38 A light elastic string has natural length l and modulus $5mg$. One end of the string is attached to a fixed point A. A particle P of mass m hangs freely in equilibrium attached to the other end of the string.

(a) Show that P is a distance $\dfrac{6l}{5}$ vertically below A.

P is pulled down a vertical distance a below its equilibrium position and released from rest at time $t = 0$.

(b) Show that provided the string remains taut

$$\frac{d^2 x}{dt^2} = -\frac{5gx}{l}$$

where x is the displacement of P from its equilibrium position.

(c) Given that $a = \dfrac{l}{10}$ state the amplitude and find, in terms of l and g, the period of the resulting oscillations.

(d) If instead $a = \dfrac{l}{4}$, show that the string first becomes slack at time $\sqrt{\left(\dfrac{5l}{4g}\right)}$.

39 A particle of mass 0.4 kg is tied to one end of a taut light inextensible string of length 0.5 m. The other end of the string is tied to a fixed point A on a smooth horizontal plane. Given that the particle moves with speed $8\,\text{m s}^{-1}$ on the plane and in a circular path with centre A, find, in N, the tension in the string. Given that the string snaps when its tension exceeds 60 N, find, in radians per second to one decimal place, the greatest angular speed at which the string can rotate.   [L]

40 A particle P, moving on the smooth inside surface of a fixed spherical bowl, of radius r, describes a horizontal circle at depth $r/2$ below the centre of the bowl. Show that P takes time $\pi\sqrt{(2r/g)}$ to complete one revolution of its circular path.   [L]

41 One end of a light inextensible string, of length a, is tied to a fixed point O and a particle Q, of mass m, is tied at the other end. The particle moves with constant angular speed ω in a horizontal circle so that the string traces out the surface of a cone. The centre of the circle is C, where C is vertically below O and angle COQ is β.

(a) Find, in terms of m, g and β, the magnitude of the tension in the string.

(b) Find ω^2 in terms of a, g and β, and hence deduce that

$$a\omega^2 > g.$$ [L]

42 A light inextensible string, of length 2 m, passes through a small fixed smooth ring R and carries particle A, of mass $2M$, at one end and particle B, of mass M, at the other end. The particle A is at rest vertically below R. The particle B moves in a horizontal circle, the centre of which is vertically below R, with

constant angular speed $4\,\text{rad}\,\text{s}^{-1}$. Given that the string remains taut, calculate the distance AR.   [L]

43 A particle is attached at one end of a light string, the other end of which is fixed. When the particle moves in a horizontal circle with speed $2\,\text{m}\,\text{s}^{-1}$, the string makes an angle $\arctan(5/12)$ with the vertical. Show that the length of the string is approximately $2.5\,\text{m}$.   [L]

44 An elastic string of length l and modulus $3mg$ has one end fixed to a point O. A particle of mass m is attached to the other end of the string and hangs in equilibrium. Find the stretched length of the string.

The particle is set in motion and describes a horizontal circle with constant angular speed ω. Given that the centre of the circle is at a distance l vertically below O, show

(a) that the stretched length of the string is $3l/2$.

(b) that $\omega^2 l = g$.   [L]

45 A particle, of mass m, is suspended from a fixed point O by a light elastic string, of natural length l and modulus λ. The particle moves with constant angular speed ω in a horizontal circular path with the string making a constant angle θ with the downward direction of the vertical. Show that x, the extension of the string, is given by

$$\lambda x = \omega^2 lm(l + x).$$

Deduce that the motion described cannot take place unless $\omega^2 < \lambda/(lm)$.

Show further that, for a given value of ω, the depth of the horizontal circle below O is independent of λ.   [L]

46 (i) A bead, of mass m, threaded on a fixed, smooth, horizontal circular hoop, of radius d, is moving with constant speed v. Given that the force exerted by the hoop on the bead is of magnitude R, and that this force acts at an angle θ to the horizontal, express

(a) R^2 in terms of v, d, m and g,

(b) $\tan\theta$ in terms of v, d and g.   [L]

(ii) State the model you used for the bead in this question. The bead is said to be 'moving with constant speed v'. For this to happen a physical force in addition to the frictional force has to be assumed negligible. Describe this force in words.

47 A parcel is placed on the rough horizontal floor of a van. The coefficient of friction between the parcel and the floor is $\frac{3}{8}$. The van rounds a horizontal circular bend of radius 100 m. Find the maximum constant speed, in metres per second, to 3 significant figures, at which the van may travel in order that the parcel shall not slide. [L]

48 A light elastic string has natural length l and modulus of elasticity $2mg$. One end of the string is attached to a fixed point A. A particle P, of mass m, is attached to the other end of the string.

(a) When P hangs vertically below A in equilibrium, find the length of the string in terms of l.

The point C is at distance l vertically below A. The particle P moves at constant speed in a horizontal circle, centre C. The string AP is inclined at an acute angle θ to AC.

(b) Show that $\theta = 60°$.

(c) Find the tension in the string, giving your answer in terms of m and g.

(d) Find the time taken for P to move once around the circle, giving your answer in terms of π, l and g. [L]

49 A particle of mass m is suspended from a fixed point A by a light inextensible string of length l. The particle moves in a horizontal circle, whose centre is vertically below A, with constant angular speed ω and with the string taut and inclined at an angle θ to the downward vertical through A.

(a) Show that $\omega^2 l \cos \theta = g$.

The string is replaced by a light elastic string of natural length a and modulus of elasticity λ. The particle now moves in a horizontal circle, whose centre is vertically below A, with constant angular speed 2ω. The elastic string makes the same angle θ with the downward vertical.

(b) Show that $\dfrac{1}{4a} - \dfrac{1}{4l} = \dfrac{m\omega^2}{\lambda}$. [L]

50 An artificial satellite of mass m moves under the action of a gravitational force which is directed towards the centre, O, of the earth and is of magnitude F. The orbit of the satellite is a circle of radius a and centre O. Obtain an expression for T, the period of the satellite, in terms of m, a and F.

Show that, if the gravitational force acting on a body of mass m at a distance r from O is $m\mu/r^2$, where μ is a constant, then $T^2\mu = 4\pi^2 a^3$.

Assuming that the radius of the earth is $6400\,\text{km}$ and that the acceleration due to gravity at the surface of the earth is $10\,\text{m s}^{-2}$, show that $\mu = (6.4)^2 10^{15}\,\text{m}^3\,\text{s}^{-2}$.

Hence, or otherwise, find the period of revolution, in hours to 2 decimal places, of the satellite when it travels in a circular orbit $600\,\text{km}$ above the surface of the earth.   [L]

The work–energy principle

4

In Book M1, section 5.3, the relationship between work and energy was studied. This relationship will now be studied in greater depth by considering more complicated situations and including other kinds of energy.

4.1 The motion of a particle in a vertical circle

A particle can be caused to move on a path which is part of a circle in a vertical plane in a variety of ways. For example, it could be attached to the end of a string or rod and set in motion, or it could be placed on top of a cylinder whose axis is horizontal and then disturbed so that it slides. For the circular motion in a horizontal plane studied in chapter 3, the angular speed was constant. For motion in a vertical circle this is not the case, so the first step is to investigate the acceleration of the particle once more.

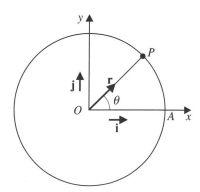

Let the particle P, which is constrained to move in a vertical plane on a circular path centre O of radius r, have position vector $\mathbf{r}$ relative to O.

Let the angle between $\mathbf{r}$ and OA, a fixed line through O, be θ at time t.

Taking unit vectors $\mathbf{i}$ and $\mathbf{j}$ along and perpendicular to OA gives:

$$\mathbf{r} = r\cos\theta\,\mathbf{i} + r\sin\theta\,\mathbf{j}$$

Differentiating with respect to *time*, as shown in Book P2, chapter 8, gives:

$$\mathbf{v} = \dot{\mathbf{r}} = -r\sin\theta\dot\theta\,\mathbf{i} + r\cos\theta\dot\theta\,\mathbf{j}$$

Differentiating again with respect to time, using the product rule for differentiation (Book P2, chapter 8):

$$\mathbf{a} = \ddot{\mathbf{r}} = (-r\cos\theta\dot\theta^2 - r\sin\theta\ddot\theta)\mathbf{i} + (-r\sin\theta\dot\theta^2 + r\cos\theta\ddot\theta)\mathbf{j}$$

This can be rearranged to give:

$$\ddot{\mathbf{r}} = -(r\cos\theta\,\mathbf{i} + r\sin\theta\,\mathbf{j})\dot\theta^2 + (r\sin\theta\,\mathbf{i} - r\cos\theta\,\mathbf{j})\ddot\theta$$
$$= -\mathbf{r}\dot\theta^2 + (r\sin\theta\,\mathbf{i} - r\cos\theta\,\mathbf{j})\ddot\theta$$

There are two separate parts to this acceleration:

(1) $-\mathbf{r}\dot\theta^2$ which is a vector of magnitude $r\dot\theta^2$ directed along PO (this is equal to $-\omega^2 r$ when $\dot\theta$ is constant, as in chapter 3).
(2) $(r\sin\theta\,\mathbf{i} - r\cos\theta\,\mathbf{j})\ddot\theta$
    (this is zero when $\dot\theta = \omega = $ constant).

Now:
$$|(r\sin\theta\,\mathbf{i} - r\cos\theta\,\mathbf{j})\ddot\theta| = |\sin\theta\,\mathbf{i} - \cos\theta\,\mathbf{j}|r\ddot\theta$$
$$= r\ddot\theta\sqrt{(\sin^2\theta + \cos^2\theta)}$$
$$= r\ddot\theta \quad \text{as } \sin^2\theta + \cos^2\theta = 1$$

So the magnitude of $(r\sin\theta\,\mathbf{i} - r\cos\theta\,\mathbf{j})\ddot\theta$ is $r\ddot\theta$.

To find its direction:

The gradient of the line representing $(r\sin\theta\,\mathbf{i} - r\cos\theta\,\mathbf{j})\ddot\theta$ is:

$$-\frac{r\cos\theta}{r\sin\theta} = -\frac{\cos\theta}{\sin\theta}$$

and the gradient of the line representing $\mathbf{r}$ is:

$$\frac{r\sin\theta}{r\cos\theta} = \frac{\sin\theta}{\cos\theta}$$

Since
$$\frac{\sin\theta}{\cos\theta} \times -\frac{\cos\theta}{\sin\theta} = -1$$

these two lines are perpendicular as shown in Book P2, chapter 2.

So $r\sin\theta\,\mathbf{i} - r\cos\theta\,\mathbf{j}$ is perpendicular to $\mathbf{r}$ and is directed along the tangent to the circular path. Hence the acceleration has two components:

$r\dot\theta^2$ along the inward radius towards O

and $r\ddot\theta$ along the tangent to the circular path.

$\dot{\theta}$ is the rate of change of θ with respect to time, that is, $\dot{\theta}$ is the angular speed of OP.

Writing this angular speed as ω, but remembering that for motion in a vertical circle ω *is not constant* gives:

$$\dot{\theta} = \omega = \frac{v}{r}$$

where v is the linear speed of P.

And so:
$$r\dot{\theta} = v$$

Differentiating with respect to time gives:

$$r\ddot{\theta} = \frac{\mathrm{d}v}{\mathrm{d}t} \quad \text{as } r \text{ is constant.}$$

The components of the acceleration are thus:

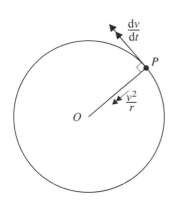

■ $\qquad r\omega^2$ **or** $\dfrac{v^2}{r}$ **along the inward radius**

and $\qquad \dfrac{\mathrm{d}v}{\mathrm{d}t} \qquad$ **along the tangent.**

In some cases of motion in a vertical circle, the particle, once in motion, is subject only to gravity and a force along the radius of the circular path from, for example, the tension in the string or rod. As this radial force is perpendicular to the direction of motion of the particle it does no work. It follows that the sum of the kinetic energy and potential energy of the particle is constant throughout the motion. This is an application of the **work–energy principle** which states that:

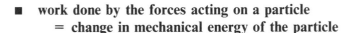

■ **work done by the forces acting on a particle**
    = change in mechanical energy of the particle

This fact is very important in solving problems as it enables you to calculate the speed of the particle at a given point and hence also the radial acceleration and the force acting on the particle.

Motion of a particle on a fixed vertical circle

A particle which is attached to one end of a rod or is threaded on a circular wire and caused to move in a vertical circle cannot leave that circular path. In some cases the particle will describe complete circles but in other cases it will move on an arc of a circle.

Consider a particle P of mass m which is attached to one end of a light rod of length r. The rod is free to rotate in a vertical plane about an axis through the other end O. Suppose P is hanging in equilibrium and is then given a speed u.

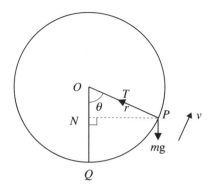

Let the speed of P be v when OP makes an angle θ with the downward vertical.

By the work–energy principle:

as the tension is perpendicular to the direction of motion and hence does no work,

$$\text{potential energy gained } = \text{ kinetic energy lost}$$

The particle has risen a vertical distance $OQ - ON$.

As $ON = r\cos\theta$, the distance $OQ - ON$ is $r - r\cos\theta$.

So: $$\text{P.E. gained} = mgh = mg(r - r\cos\theta)$$

And: $$\text{K.E. lost} = \tfrac{1}{2}mu^2 - \tfrac{1}{2}mv^2$$

Hence: $$mg(r - r\cos\theta) = \tfrac{1}{2}mu^2 - \tfrac{1}{2}mv^2$$

Rearranging this gives:

$$\tfrac{1}{2}mv^2 = \tfrac{1}{2}mu^2 - mg(r - r\cos\theta)$$

And so: $$v^2 = u^2 - 2gr(1 - \cos\theta) \qquad (1)$$

If P is to describe complete circles, it must have a speed at the highest point, that is $v^2 > 0$ when $\theta = 180°$.

When $\theta = 180°$: $$v^2 = u^2 - 2gr(1 - \cos 180°)$$
$$v^2 = u^2 - 2gr(1 - (-1))$$
$$v^2 = u^2 - 4gr$$

And for $v^2 > 0$: $$u^2 - 4gr > 0$$
$$u^2 > 4gr$$

So a particle moving on a fixed vertical circle of radius r will move in complete circles if its speed u at the lowest point satisfies

$$u^2 > 4gr$$

If this is not the case, that is if $u^2 - 4gr < 0$, the speed must become zero before the particle reaches the highest point of the circle and the particle will oscillate on the arc ABC of the circle as shown in the diagram.

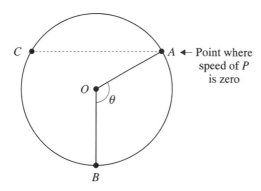

From equation (1) above, the speed v at any point on the path is given by:

$$v^2 = u^2 - 2gr(1 - \cos\theta)$$

and substituting $v = 0$ enables you to calculate the maximum value of θ:

$$0 = u^2 - 2gr(1 - \cos\theta_{max})$$
$$0 = u^2 - 2gr + 2gr\cos\theta_{max}$$
$$\cos\theta_{max} = \frac{2gr - u^2}{2gr}$$
$$\theta_{max} = \arccos\left(\frac{2gr - u^2}{2gr}\right)$$

Example 1

A particle of mass 0.5 kg is attached to one end A of a light rod AB of length 0.2 m which is free to rotate in a vertical plane about the other end B. The particle is held at rest vertically above B. It is slightly displaced so that it moves in a vertical circle.

Calculate:

(a) the speed of the particle as it passes through the lowest point of its path

(b) the tension in the rod at this point.

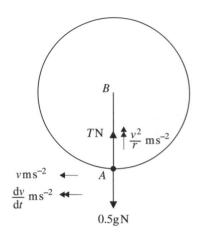

(a) Let the speed of the particle at its lowest point be $v\,\mathrm{m\,s^{-1}}$ and the tension in the rod be T N.

At the lowest point the particle has moved a distance of 0.4 m downwards from its initial position.

So: potential energy lost $= mgh = 0.5 \times g \times 0.4\,\mathrm{J}$

and: kinetic energy gained $= \frac{1}{2}mv^2 - 0 = \frac{1}{2} \times 0.5v^2\,\mathrm{J}$

Using the work–energy principle:

$$\text{K.E. gained} = \text{P.E. lost}$$

So: $\frac{1}{2} \times 0.5 \times v^2 = 0.5 \times g \times 0.4$

$$v^2 = 9.8 \times 0.8$$

$$v = 2.8$$

The speed of the particle at its lowest point is $2.8\,\mathrm{m\,s^{-1}}$.

(b) To calculate the tension in the rod, write down the vertical component of the equation of motion.

Using: $F = ma$

gives: $T - 0.5\mathrm{g} = 0.5\,\dfrac{v^2}{r}$

Since $v = 2.8\,\mathrm{m\,s^{-1}}$ and $r = 0.2\,\mathrm{m}$ this gives:

$$T = 0.5 \times 9.8 + 0.5 \times \frac{2.8^2}{0.2}$$

and so: $T = 24.5$

The tension in the rod is 24.5 N.

Example 2

A bead P is threaded on a smooth circular wire of radius 0.5 m and centre O which is fixed in a vertical plane. The bead is projected from the lowest point of the wire with a speed u m s^{-1}.

(a) Describe the model you would use for the bead in order to investigate its motion.
(b) Show that if $u^2 > 19.6$ the bead will describe complete circles.
(c) Given that $u^2 = 4.9$ calculate the angle between OP and the downward vertical at the highest point of P's path.

(a) The wire has radius 0.5 m and relative to that the bead's radius is small, so the bead can be modelled as a particle.

(b) For the bead to describe complete circles, it must have a speed at the highest point of the circular path. Let this speed be v m s^{-1}. Then $v^2 > 0$ if the bead is to describe complete circles.

Using the work–energy principle:

$$\text{K.E. lost} = \text{P.E. gained}$$

$$\tfrac{1}{2}mu^2 - \tfrac{1}{2}mv^2 = mgh$$

where m kg is the mass of the bead.

But: $\qquad\qquad h = 2 \times \text{radius} = 2 \times 0.5 = 1$

So: $\qquad\qquad \tfrac{1}{2}mu^2 - \tfrac{1}{2}mv^2 = m \times 9.8 \times 1$

$$u^2 - v^2 = 9.8 \times 2$$

$$v^2 = u^2 - 19.6$$

Since: $\qquad\qquad u^2 > 19.6$

it follows that: $\qquad\qquad v^2 > 0$

And so the bead will move in complete circles.

(c) The highest point of P's path will be the point where P's speed is zero.

Let the angle between OP and the downward vertical at this point be θ.

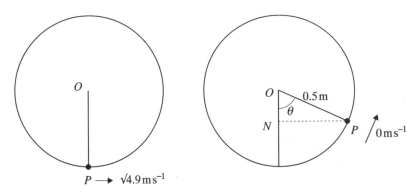

Initial situation P at the highest point

Since $ON = 0.5\cos\theta$, the vertical distance P has risen is:

$$0.5 - 0.5\cos\theta = 0.5(1 - \cos\theta)$$

Using the work–energy principle:

$$\tfrac{1}{2}\,mu^2 - 0 = mgh$$
$$\tfrac{1}{2}\,m \times 4.9 = m \times 9.8 \times 0.5(1 - \cos\theta)$$
$$4.9 = 9.8 \times (1 - \cos\theta)$$
$$\tfrac{1}{2} = 1 - \cos\theta$$
$$\cos\theta = \tfrac{1}{2}$$
$$\theta = 60°$$

The angle between OP and the downward vertical at the highest point of P's path is $60°$.

Exercise 4A

Whenever a numerical value of g is required, take $g = 9.8\,\mathrm{m\,s^{-2}}$.

1 A particle of mass 1.5 kg is attached to one end A of a light rod AB of length 0.75 m which is free to rotate in a vertical plane about the other end B. The rod is hanging with A below B. The particle is projected horizontally with speed $10\,\mathrm{m\,s^{-1}}$. Calculate (a) the speed of the particle when it is vertically above B (b) the tension in the rod at this time.

2 A particle of mass 0.7 kg is attached to one end A of a light rod AB of length 0.5 m which is free to rotate in a vertical plane about the other end B. AB is held in a horizontal position and

then released. Calculate the speed of the particle at the lowest point of its path.

3   A pendulum bob has mass 0.2 kg. It is attached to one end of a light rod of length 2 m. The rod is free to rotate in a vertical plane about an axis through the other end O. Given that the pendulum swings through 60° on either side of the vertical, calculate the speed of the bob at the lowest point of its path and the tension in the rod when the rod makes an angle of 30° with the downward vertical.

4   A particle of mass 0.5 kg is attached to one end A of a light rod AB of length 2 m which is free to rotate in a vertical plane about the other end B. AB hangs vertically with A below B. The particle is projected horizontally with speed 12 m s^{-1}.
(a) Calculate the speed of the particle when AB is horizontal. When AB is horizontal the particle strikes and adheres to an identical particle which is at rest. (b) Calculate the speed of the combined particle and determine whether this combined particle will describe complete circles.

5   A particle P is attached to one end of a light rod of length l which is free to rotate in a vertical plane about its other end. The rod hangs vertically with P below the fixed end. (a) Given that P is projected horizontally with speed $3\sqrt{(lg)}$ prove that P describes complete circles. (b) Find the least value of the speed of projection if P is to describe complete circles.

6   A bead B of mass m is threaded on a smooth circular wire of radius a and centre O which is fixed in a vertical plane. B is released from rest at the point where OB makes an angle of 30° with the upward vertical. (a) State how you could model the bead in order to investigate its motion, showing clearly any assumptions made. (b) Given that the speed of B at the lowest point of its path is v, find v^2 in terms of a and g. (c) Find in terms of m and g the magnitude and direction of the reaction of the wire on the bead when OB makes an angle of 60° with the upward vertical.

7   A particle P of mass 2 kg is attached to one end of a light rod of length 0.5 m which is free to rotate in a vertical plane about its other end. The particle describes complete circles. Given that

the tension at the lowest point of P's path is five times the tension at the highest point, find the speed of P at its lowest point.

Motion of a particle which can leave the circular path

In each of examples 1 and 2 the particle was unable to leave the circular path on which it was moving. However, if a particle P is attached to one end of a *string* and set in motion in a vertical circle, the motion of the particle will be described by one of three distinct cases.

(1)

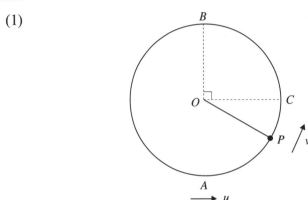

The speed of the particle at the lowest point A is sufficient for the particle to pass through the highest point B of the circular path *with the string still taut*. In this case the particle will describe complete circles.

(2)

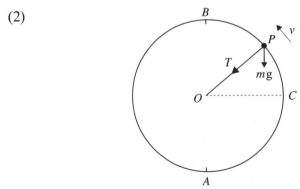

The speed of the particle at A is sufficient for the particle to pass through the point C on the same horizontal level as the centre of the circle but the string becomes slack at some point between C and B. Once the string becomes slack the only force acting on the particle is gravity, but the particle will have a velocity and so will move as a projectile until the string becomes taut once more.

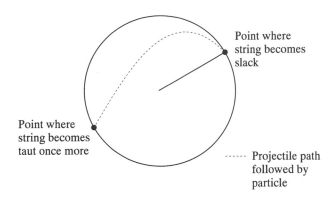

Point where string becomes slack

Point where string becomes taut once more

------ Projectile path followed by particle

(3) The speed of the particle at A is insufficient for the particle to rise above the level of the centre of the circle. In this case the particle oscillates as a pendulum.

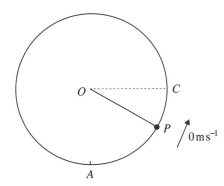

For a particle attached to a string, the motion can only be in a circle when the string is taut.

Hence:

the condition for complete circles is that there is a tension at the highest point of the circular path.

Consider a particle P of mass m attached to one end of a light inextensible string of length r. The other end of the string is fixed at a point O and P hangs freely in equilibrium. Suppose P is projected horizontally with speed u.

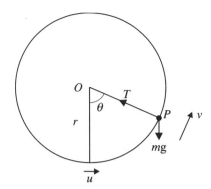

Let the speed of P when OP makes an angle θ with the downward vertical be v and the tension in the string be T.

By the work–energy principle:

$$\text{K.E. lost} = \text{P.E. gained}$$
$$\tfrac{1}{2}mu^2 - \tfrac{1}{2}mv^2 = mgh$$
$$\tfrac{1}{2}mu^2 - \tfrac{1}{2}mv^2 = mgr(1 - \cos\theta)$$
$$u^2 - v^2 = 2gr(1 - \cos\theta)$$
$$v^2 = u^2 - 2gr(1 - \cos\theta) \qquad (1)$$

Using the equation of motion, $F = ma$, along PO, with $a = \dfrac{v^2}{r}$ gives:

$$T - mg\cos\theta = m\frac{v^2}{r} \qquad (2)$$

Eliminating v^2 between equations (1) and (2) gives:

$$T - mg\cos\theta = \frac{m}{r}\left[u^2 - 2gr(1 - \cos\theta)\right]$$
$$T = mg\cos\theta + m\frac{u^2}{r} - 2mg(1 - \cos\theta)$$
$$T = m\left(\frac{u^2}{r} - 2g + 3g\cos\theta\right)$$

Case (1)   For complete circles the string must be taut at the highest point of the path.

That is: $\qquad\qquad\qquad T \geqslant 0$ when $\theta = 180°$

So: $\qquad\qquad m\left(\dfrac{u^2}{r} - 2g + 3g\cos 180°\right) \geqslant 0$

$$\frac{u^2}{r} \geqslant 2g - 3g\cos 180°$$

$$\frac{u^2}{r} \geqslant 2g + 3g \qquad \text{as } \cos 180° = -1$$

$$u^2 \geqslant 5rg$$

Case (2)   The string becomes slack when the particle is above the level of *O*.

That is: $T = 0$ for some θ between 90° and 180°

From above: $T = m\left(\dfrac{u^2}{r} - 2g + 3g\cos\theta\right)$

So $T = 0$ when: $3g\cos\theta = 2g - \dfrac{u^2}{r}$

that is, when: $\cos\theta = \dfrac{2rg - u^2}{3rg}$

But: $90° < \theta < 180° \Rightarrow 0 > \cos\theta > -1$

So: $0 > \dfrac{2rg - u^2}{3rg} > -1$

Separating this into two inequalities gives:

$$0 > \frac{2rg - u^2}{3rg} \qquad (3)$$

and: $$\frac{2rg - u^2}{3rg} > -1 \qquad (4)$$

As $3rg$ must be positive you can multiply these inequalities by $3rg$.

So from (3): $u^2 > 2rg$

and from (4): $2rg - u^2 > -3rg$

$5rg > u^2$

Hence the string becomes slack when the particle is above the level of *O* if:

$$2rg < u^2 < 5rg$$

Case (3)   The particle oscillates like a pendulum if the speed is zero before the particle reaches the level of *O*.

That is: $v = 0$ for some $\theta \leqslant 90°$

From equation (1) above:

$$v^2 = u^2 - 2gr(1 - \cos\theta)$$

So $v = 0$ when: $u^2 - 2gr + 2gr\cos\theta = 0$

that is, when: $\cos\theta = \dfrac{2gr - u^2}{2gr}$

But as $0 < \theta \leqslant 90°$ it follows that $\cos\theta \geqslant 0$.

So: $u^2 \leqslant 2gr$

Hence, if the speed of the particle at the lowest point of its path is u:

$u^2 \geqslant 5gr \Rightarrow$ the particle will describe complete circles

$2gr < u^2 < 5gr \Rightarrow$ the particle will rise above the level of the centre of the circle and will leave the circular path

$u^2 \leqslant 2gr \Rightarrow$ the particle will oscillate as a pendulum.

Example 3

A particle of mass m is attached to one end A of a light inelastic string of length r, the other end B being fixed. Initially the particle hangs vertically below B. It is projected horizontally with speed $2\sqrt{(gr)}$. Show that the particle leaves its circular path when AB makes an angle $\arccos \frac{2}{3}$ with the upward vertical.

The particle will leave its circular path when the string becomes slack.

Let the angle between AB and the upward vertical at this moment be θ. Let the speed of the particle be v and the tension in the string be T.

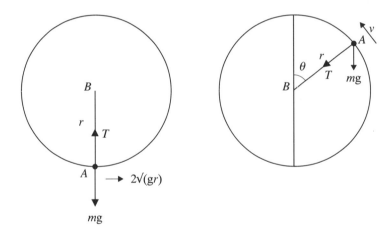

The component of the equation of motion along the inward radius gives:

$$T + mg\cos\theta = m\,\frac{v^2}{r}$$

When $T = 0$:

$$mg\cos\theta = m\frac{v^2}{r}$$

$$\cos\theta = \frac{v^2}{gr} \qquad\qquad (1)$$

Using the work–energy principle gives:

$$\tfrac{1}{2}\,mu^2 - \tfrac{1}{2}\,mv^2 = mgr(1 + \cos\theta)$$

As $u = 2\sqrt{(gr)}$:

$$\tfrac{1}{2}m \times 4gr - \tfrac{1}{2}\,mv^2 = mgr + mgr\cos\theta$$

$$4gr - v^2 = 2gr + 2gr\cos\theta$$

$$v^2 = 2gr - 2gr\cos\theta$$

$$\frac{v^2}{gr} = 2 - 2\cos\theta$$

But from equation (1) above $\dfrac{v^2}{gr} = \cos\theta$

So:

$$\cos\theta = 2 - 2\cos\theta$$

$$3\cos\theta = 2$$

$$\cos\theta = \tfrac{2}{3}$$

$$\theta = \arccos\tfrac{2}{3}$$

The particle will leave its circular path when AB makes an angle $\arccos\tfrac{2}{3}$ with the upward vertical.

A particle which moves on a circular cylinder or sphere, either on the inside or outside of it, will experience a contact force from the cylinder or sphere. When that contact force becomes zero the particle will leave its circular path.

Example 4

A particle P of mass 2 kg rests at the highest point of a smooth sphere centre O of radius 0.5 m which is fixed to a horizontal table. P is slightly disturbed and slides on the surface of the sphere.

(a) Determine the angle between OP and the upward vertical when the particle leaves the surface of the sphere.
(b) Calculate the magnitude and direction of the particle's velocity when it hits the table.

(a) Consider the particle when OP has turned through an angle θ. Let the speed of the particle at this moment be $v\,\mathrm{m\,s^{-1}}$. Let the normal contact force between P and the sphere be R N.

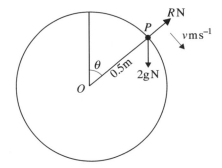

Using the work–energy principle gives:

$$\tfrac{1}{2}mv^2 - \tfrac{1}{2}mu^2 = mgh$$
$$\tfrac{1}{2} \times 2 \times v^2 - 0 = 2 \times 9.8 \times 0.5(1 - \cos\theta)$$
$$v^2 = 9.8(1 - \cos\theta) \qquad (1)$$

Using the equation of motion, $F = m\frac{v^2}{r}$, along the inward radius gives:

$$2 \times 9.8\cos\theta - R = \frac{2v^2}{0.5}$$

$$v^2 = \tfrac{1}{4}(2 \times 9.8\cos\theta - R) \qquad (2)$$

Equating the expressions for v^2 from equations (1) and (2) gives:

$$\tfrac{1}{4}(2 \times 9.8\cos\theta - R) = 9.8(1 - \cos\theta)$$

The particle leaves the surface when $R = 0$, that is when:

$$\tfrac{1}{2} \times 9.8\cos\theta = 9.8(1 - \cos\theta)$$

$$\tfrac{1}{2}\cos\theta = 1 - \cos\theta$$

$$\tfrac{3}{2}\cos\theta = 1$$

$$\cos\theta = \tfrac{2}{3}$$

$$\theta = 48.18°$$

The particle leaves the surface when the angle between OP and the upward vertical is 48.2°.

(b) Once the particle leaves the surface of the sphere the only force acting upon it is gravity. The particle will therefore move as a projectile (see Book M1, section 3.5).

To investigate the motion of this projectile you must first find the velocity of the particle as it leaves the sphere.

From equation (2) above:

$$v^2 = \tfrac{1}{4}(2 \times 9.8 \times \cos\theta - R)$$

When the particle leaves the sphere, $\cos\theta = \tfrac{2}{3}$ and $R = 0$.

So:
$$v^2 = \tfrac{1}{4} \times 2 \times 9.8 \times \tfrac{2}{3}$$
$$v^2 = \frac{9.8}{3}$$

OP makes an angle of $48.18°$ with the upward vertical at the moment the particle leaves the sphere.

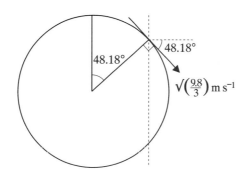

Using geometry, the diagram shows that the particle's velocity is at an angle of $48.18°$ below the horizontal.

So for the projectile motion, the particle's velocity has components:

$$\sqrt{\left(\frac{9.8}{3}\right)} \cos 48.18° \quad \text{horizontally}$$

and:
$$\sqrt{\left(\frac{9.8}{3}\right)} \sin 48.18° \quad \text{vertically downwards}$$

The vertical distance travelled by the particle from leaving the sphere to reaching the table is:

$$r(1 + \cos 48.18°) = 0.5 \times \frac{5}{3} = \frac{2.5}{3} \text{ m}$$

as $\cos 48.18° = \tfrac{2}{3}$ from part (a).

For the vertical speed, using $v^2 = u^2 + 2as$ with $a = 9.8\,\text{m s}^{-2}$,
$u = \sqrt{\left(\frac{9.8}{3}\right)} \sin 48.18°\,\text{m s}^{-1}$ and $s = \frac{2.5}{3}\,\text{m}$ gives:

$$v^2 = \frac{9.8}{3}\sin^2 48.18° + 2 \times 9.8 \times \frac{2.5}{3}$$
$$v = 4.259\,\text{m s}^{-1}$$

There is no horizontal acceleration and so the horizontal speed is unchanged throughout the projectile motion.

Horizontal speed $= \sqrt{\left(\dfrac{9.8}{3}\right)} \cos 48.18°$

$= 1.205 \,\mathrm{m\,s^{-1}}$

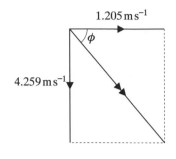

From the vector diagram, the resultant speed is:

$$\sqrt{\{1.205^2 + 4.259^2\}} = 4.426 \,\mathrm{m\,s^{-1}}$$

and: $\tan \phi = \dfrac{4.259}{1.205}$

so: $\phi = 74.20°$

The particle hits the table with a speed of $4.43 \,\mathrm{m\,s^{-1}}$ at an angle of $74.2°$ below the horizontal.

Example 5

A smooth hemispherical bowl of radius r and centre O is fixed with its rim horizontal. A marble P of mass m is projected vertically downwards inside the bowl from a point on the rim. The initial speed of P is u. When P reaches the lowest point of the bowl it strikes a stationary marble Q of mass $2m$. P is brought to rest by the impact.

  (a) Describe the models you would use for P and Q to investigate the subsequent motion, stating clearly any assumptions made.
  (b) Show that Q will leave the bowl provided

$$u^2 > 6gr$$

  (c) Given that $u^2 = 4gr$, calculate the magnitude of the reaction between Q and the bowl at the highest point of Q's path.

(a) A marble can be assumed to have a small radius compared with the radius of the bowl. P and Q can therefore be modelled as particles.

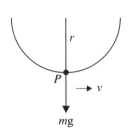

As the bowl is described as being smooth, it can be assumed that there is no frictional force acting on the marbles.

(b) Consider the motion of P from the rim to the lowest point of the bowl. Let the speed of P at the lowest point be v.

$$\text{K.E. gained by } P = \tfrac{1}{2}mv^2 - \tfrac{1}{2}mu^2$$
$$\text{P.E. lost by } P = mgh = mgr$$

By the work–energy principle:

$$\text{K.E. gained} = \text{P.E. lost}$$
$$\tfrac{1}{2}mv^2 - \tfrac{1}{2}mu^2 = mgr$$
$$v^2 = 2gr + u^2$$

Consider the impact of P and Q.

Before impact:

$\longrightarrow \sqrt{(2gr + u^2)}$ $\longrightarrow 0$

$\underset{m}{\underset{\bigcirc}{P}}$ $\underset{2m}{\underset{\bigcirc}{Q}}$

After impact:

$\longrightarrow 0$ $\longrightarrow V$

$\underset{m}{\underset{\bigcirc}{P}}$ $\underset{2m}{\underset{\bigcirc}{Q}}$

Let the speed of Q after impact be V.

By the principle of conservation of momentum (see Book M1, section 5.4):

$$m_1 u_1 + m_2 u_2 = m_1 v_1 + m_2 v_2$$

So:

$$m\sqrt{(2gr + u^2)} + 0 = 0 + 2mV$$
$$V = \tfrac{1}{2}\sqrt{(2gr + u^2)}$$

Consider Q when OQ makes an angle θ with the downward vertical. Let the speed of Q at this point be W.

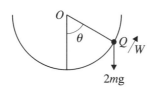

From the lowest point to this level:

$$\text{P.E. gained by } Q = mgh = 2mgr(1 - \cos\theta)$$
$$\text{K.E. lost by } Q = \tfrac{1}{2} \times 2mV^2 - \tfrac{1}{2} \times 2mW^2$$
$$= \tfrac{1}{4}m(2gr + u^2) - mW^2$$

as $V = \tfrac{1}{2}\sqrt{(2gr + u^2)}$.

By the work–energy principle:

$$\text{K.E. lost} = \text{P.E. gained}$$

$$\tfrac{1}{4}m(2gr + u^2) - mW^2 = 2mgr(1 - \cos\theta)$$

$$\tfrac{1}{2}gr + \tfrac{1}{4}u^2 - W^2 = 2gr - 2gr\cos\theta \qquad (1)$$

$$W^2 = \tfrac{1}{4}u^2 - \tfrac{3}{2}gr + 2gr\cos\theta$$

Q will leave the bowl provided $W^2 > 0$ when $\theta = 90°$, that is provided:

$$\tfrac{1}{4}u^2 - \tfrac{3}{2}gr + 0 > 0$$

or: $$u^2 > 6gr$$

(c) From equation (1) above:

$$\tfrac{1}{2}gr + \tfrac{1}{4}u^2 - W^2 = 2gr - 2gr\cos\theta$$

As $u^2 = 4gr$ this gives:

$$\tfrac{1}{2}gr + gr - W^2 = 2gr - 2gr\cos\theta$$

At the highest point of Q's path $W = 0$.

So: $$\tfrac{1}{2}gr + gr = 2gr - 2gr\cos\theta$$

$$2gr\cos\theta = \tfrac{1}{2}gr$$

$$\cos\theta = \tfrac{1}{4}$$

Let the normal reaction between Q and the bowl be R.

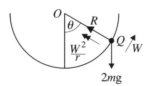

Using Newton's second law, $F = ma$, along the radius with $a = \dfrac{W^2}{r}$ gives:

$$R - 2mg\cos\theta = 2m\frac{W^2}{r}$$

but at the highest point of Q's path, $W = 0$ and $\cos\theta = \tfrac{1}{4}$

so:

$$R - 2mg \times \tfrac{1}{4} = 0$$

$$R = \tfrac{1}{2}mg$$

The magnitude of the normal reaction between Q and the bowl is $\tfrac{1}{2}mg$.

Exercise 4B

Whenever a numerical value of g is required, take $g = 9.8\,\mathrm{m\,s^{-2}}$.

1 A light inelastic string of length 1 m has one end fixed to a point O. A particle P of mass $0.25\,\mathrm{kg}$ is hanging in equilibrium from the free end of the string. P is projected horizontally with a speed which is just sufficient for P to reach the level of O. Calculate the speed of projection and the tension in the string when OP makes an angle of $60°$ with the vertical.

2 A light inelastic string of length 2 m has one end fixed to a point O. A particle P of mass $1\,\mathrm{kg}$ is hanging in equilibrium from the free end of the string. P is projected horizontally.
(a) Calculate the minimum speed of projection if P is to describe complete circles.
(b) Given that P is projected at twice this minimum speed, calculate the maximum and minimum magnitudes of the tension in the string.

3 A particle P of mass $0.5\,\mathrm{kg}$ is hanging in equilibrium from one end of a light inelastic string of length $1.5\,\mathrm{m}$ whose other end is fixed at a point O. P is projected horizontally with speed $u\,\mathrm{m\,s^{-1}}$. (a) Given that the string becomes slack when OP makes an angle of $45°$ with the upward vertical, calculate the value of u. (b) Describe briefly the subsequent motion of the particle.

4 A particle P of mass $0.5\,\mathrm{kg}$ is hanging in equilibrium from one end of a light inelastic string of length $1.5\,\mathrm{m}$ whose other end is fixed at a point O. P is projected horizontally with speed $u\,\mathrm{m\,s^{-1}}$. When OP is horizontal it meets a small smooth peg at Q where $OQ = 1\,\mathrm{m}$.
(a) Calculate the minimum value of u if P is to describe a complete circle about Q.
(b) Given that P is projected with $\frac{9}{10}$ of this speed, calculate the angle between PQ and the vertical when the string becomes slack.

5 A child of mass $20\,\mathrm{kg}$ is sitting on a swing which has ropes of length $2.5\,\mathrm{m}$. She is pulled back until the ropes make an angle of $50°$ with the downward vertical and then released.

(a) State how you would model the child in order to investigate the subsequent motion.

(b) Calculate the speed with which the child passes through the lowest point of the swing and calculate the tension in each of the two ropes at this point.

6 A light inelastic string of length l has one end fixed to a fixed point O. A particle P of mass m is attached to the other end. P is held vertically above O with the string taut and is projected horizontally with speed $\sqrt{(gl)}$.

(a) Find, in terms of l, m, g and θ the tension in the string when OP makes an angle θ with the upward vertical.

(b) Given that the string will break when the tension is $4mg$, find the angle between OP and the upward vertical when the string breaks.

7 A particle of mass m is attached to end A of a light inelastic string AB of length l which is fixed at B. The particle is held with AB horizontal and is projected vertically with speed u. Show that the least value of u for the particle to describe complete circles is independent of the direction of projection. Calculate this least value.

8 A particle P of mass m is at rest at the lowest point of a smooth hemispherical bowl of radius r and centre O whose rim is horizontal. P is projected horizontally with speed u.

(a) If P oscillates through $60°$ on either side of the vertical, find u in terms of r and g and find the magnitude of the reaction between P and the bowl at the highest point of P's path.

(b) If P is about to lose contact with the bowl when OP is horizontal, find u in terms of r and g.

9 A smooth hemispherical bowl of radius a has its rim horizontal. A particle P is projected downwards with speed u from a point on the surface of the bowl $\frac{1}{2}a$ below the level of the rim. If P remains in contact with the bowl during the subsequent motion show that $u \leqslant \sqrt{(ag)}$.

If $u = \sqrt{(2ag)}$, find the speed of P as it leaves the bowl.

10 A particle P rests at the highest point of a fixed smooth sphere of radius a and centre O. P is projected horizontally with speed

u. Given that P leaves the surface of the sphere when OP makes an angle θ with the upward vertical, show that:

$$\cos \theta = \frac{2ga + u^2}{3ga}$$

Hence show that if $u \geqslant \sqrt{(ag)}$ the particle will leave the surface of the sphere the instant it is projected.

11 A smooth hemispherical bowl with centre O and radius a is fixed on a horizontal surface so that its rim is in a vertical plane. A small bead is projected horizontally with speed u from the lowest point A of the rim so that it moves in a vertical circle with centre O. When the bead reaches the highest point B of the rim it is on the point of losing contact with the bowl. (a) State the mathematical model you could use for the bead in order to investigate its motion. (b) Find the speed of the bead at B in terms of a and g. (c) Find u in terms of a and g.
The bead strikes the horizontal surface again at C. (d) Find the distance AC. (e) State an assumption you have made about the forces acting during this motion.

12 A bowl is made from a smooth spherical shell of radius r by cutting away the part which is more than $\frac{1}{3}r$ above the horizontal plane through its centre. A marble is projected from the lowest point of the bowl with speed u. Show that in the subsequent motion the marble will leave the bowl and not fall back into it provided:

$$u^2 > \frac{17gr}{3}$$

4.2 Energy stored in an elastic string or spring

Consider a particle attached to one end of an elastic string whose other end is fixed on a smooth horizontal table. If the string is stretched beyond its natural length by pulling the particle along the table and then the particle is released, the particle will be seen to move on the table. That is, it will acquire kinetic energy. As the motion is horizontal there can be no change in the gravitational potential energy so it follows that when stretched, the string has energy stored in it. This form of potential energy is called the **elastic potential energy** (E.P.E.) of the string.

By the work–energy principle this elastic potential energy must be equal to the work done in stretching the string, that is the work done against the tension in the string. By Hooke's law, the tension varies with the extension in the string, $T = \dfrac{\lambda x}{l}$. A graph of the tension against the extension is a straight line:

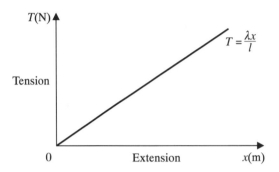

Hence it is not possible to calculate the work done against the tension by simply multiplying the tension by the extension. Instead an integration method, similar to that used to find the area under a graph (as in Book P1, Chapter 9) must be used.

First consider the work done in extending the string from an extension s to an extension $s + \delta s$ where δs is a very small length. The tension may be considered to be constant over this very small distance.

Let the tension when the extension is s be T_s.

Using: work done = force × distance moved

Gives: work done = $T_s \times \delta s$

From Hooke's law: $T_s = \dfrac{\lambda s}{l}$

and so: work done = $\left(\dfrac{\lambda s}{l}\right)\delta s$

The total work done in stretching the string from its natural length (where $s = 0$) to a length $l + x$ (where $s = x$) is given by adding the individual quantities of work for each extension δs together.

Hence:
$$\text{work done} = \sum_{s=0}^{x} \frac{\lambda s \delta s}{l}$$

Now let δs become even smaller, that is let $\delta s \to 0$.

$$\begin{aligned}
\text{Work done} &= \lim_{\delta s \to 0} \sum_{s=0}^{x} \frac{\lambda s \delta s}{l} \\
&= \int_{0}^{x} \frac{\lambda s}{l} \, \mathrm{d}s \\
&= \left[\frac{\lambda s^2}{2l} \right]_{0}^{x} \\
&= \frac{\lambda x^2}{2l}
\end{aligned}$$

- **The work done in stretching an elastic spring with modulus λ from its natural length l to a length $(l + x)$ is $\dfrac{\lambda x^2}{2l}$**

When λ is in newtons and x and l are in metres, the work done is in joules.

By the work–energy principle, the E.P.E. stored in a string extended by a length x beyond its natural length l is equal to the work done in stretching the string, that is $\dfrac{\lambda x^2}{2l}$

The elastic potential energy stored in a spring compressed by a length x from its natural length l is also given by $\dfrac{\lambda x^2}{2l}$

A string (or spring) which has its extension (or compression) increased from x_1 to x_2 will have its elastic energy increased. This increase is given by:

$$\begin{aligned}
\text{increase in E.P.E.} &= \text{final E.P.E.} - \text{ initial E.P.E.} \\
&= \frac{\lambda x_2^2}{2l} - \frac{\lambda x_1^2}{2l}
\end{aligned}$$

Example 6

An elastic string has natural length 2 m and modulus 15 N. Calculate the energy stored in the string when the string is stretched to a length of 2.6 m.

The energy stored in the string (E.P.E.) $= \dfrac{\lambda x^2}{2l}$

Since $\lambda = 15 \, \text{N}$, $l = 2 \, \text{m}$ and $x = 2.6 - 2 = 0.6 \, \text{m}$

$$\text{E.P.E.} = \frac{15 \times 0.6^2}{2 \times 2} = 1.35$$

The energy stored in the string is 1.35 J.

Example 7

A spring of natural length 0.75 m and modulus 80 N is initially compressed to a length of 0.7 m. It is then further compressed to a length of 0.5 m. Calculate the increase in the energy in J stored in the spring.

$$\text{Initial compression } x_1 = 0.75 - 0.7 = 0.05\,\text{m}$$

$$\text{Final compression } x_2 = 0.75 - 0.5 = 0.25\,\text{m}$$

Using:
$$\text{increase in energy} = \frac{\lambda x_2^2}{2l} - \frac{\lambda x_1^2}{2l}$$

with $\lambda = 80\,\text{N}$, $l = 0.75\,\text{m}$, $x_1 = 0.05\,\text{m}$ and $x_2 = 0.25\,\text{m}$ gives:

$$\text{increase} = \frac{80 \times 0.25^2}{2 \times 0.75} - \frac{80 \times 0.05^2}{2 \times 0.75}$$

$$= 3.2$$

The increase in the energy stored in the spring is 3.2 J.

Exercise 4C

Whenever a numerical value of g is required, take $g = 9.8\,\text{m s}^{-2}$.

1 An elastic string has natural length 0.8 m and modulus 10 N. Calculate the elastic potential energy stored in the string when the string is stretched to a length of 1.2 m.

2 An elastic spring of natural length 0.5 m and modulus 150 N is compressed to a length of 0.4 m. Calculate the elastic potential energy stored in the spring.

3 An elastic string has natural length 0.5 m and modulus 8 N. Calculate the work done in stretching the string (a) from its natural length to a length of 0.8 m (b) from a length of 0.75 m to 1.3 m.

4 An elastic spring has natural length 0.8 m and modulus 200 N. Calculate the work done in compressing the spring (a) from its natural length to a length of 0.7 m (b) from a length of 0.75 m to a length of 0.65 m.
   In each of parts (a) and (b) the spring is compressed by the same distance. What do you notice about the changes in energy?

5   A particle of mass 1.5 kg is hanging in equilibrium at the free end of a light elastic string of natural length 2 m and modulus 16 N. The other end of the string is fixed. The particle is pulled downwards until the string has length 4 m. Calculate the increase in the elastic energy stored in the string.

6   An elastic string has natural length l. One end of the string is fixed and a particle of mass m is attached to the other end. When the particle hangs in equilibrium the length of the string is $2l$. Calculate the energy stored in the string.

7   An elastic string has natural length l. One end of the string is fixed and a particle of mass $2m$ is attached to the other end. When the particle hangs in equilibrium the length of the string is $\frac{3}{2}l$. Calculate the work done in moving the particle from its equilibrium position to the point where the string has length $2l$.

8   One end of an elastic string of natural length l is fixed. When a particle of mass $2m$ hangs in equilibrium from the free end the length of the string is $2l$. Calculate the energy stored in the string when a particle of mass m hangs in equilibrium from the free end of the string.

4.3 Problems involving kinetic energy, gravitational potential energy and elastic potential energy

In Book M1 the work–energy principle was applied to problems involving kinetic energy and gravitational potential energy. This work can now be extended to include problems involving elastic strings or springs and elastic potential energy.

If a particle attached to the free end of an elastic string is held in equilibrium with the elastic string stretched and then released, the particle will move. As the extension of the string becomes less, the elastic potential energy stored in the string will also be reduced. The elastic potential energy is transferred to the particle which shows a corresponding increase in kinetic energy. If there is any vertical motion, the gravitational potential energy must also be taken into account.

By the work–energy principle, the total change of the kinetic, gravitational and elastic potential energies of the particle must be equal to the work done by any external forces acting on the particle.

Example 8

A particle P of mass 2 kg is attached to one end of a light elastic string of natural length 1.6 m and modulus 20 N. The other end of the string is fixed to a point O on a smooth horizontal table. P is released from rest at a point on the table 2.5 m from O and moves on the table. Calculate the speed of P when the string becomes slack.

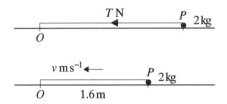

The string becomes slack when it returns to its natural length, that is when $OP = 1.6$ m.

Let the speed of P when $OP = 1.6$ m be v m s^{-1}.

When the string is slack it has no elastic potential energy.

The energy stored in the string when $OP = 2.5$ m is given by:

$$\text{elastic potential energy} = \frac{\lambda x^2}{2l}$$

where $\lambda = 20$ N, $l = 1.6$ m, $x = (2.5 - 1.6) = 0.9$ m

So:
$$\text{E.P.E.} = \frac{20 \times 0.9^2}{2 \times 1.6} \text{J}$$

$$\text{Initial kinetic energy} = 0 \text{ J}$$

$$\text{Final kinetic energy} = \tfrac{1}{2}mv^2 = \tfrac{1}{2} \times 2 \times v^2 = v^2 \text{ J}$$

By the work–energy principle:

$$\text{kinetic energy gained by particle} = \text{E.P.E. lost by string}$$

So:
$$v^2 = \frac{20 \times 0.9^2}{2 \times 1.6}$$

$$v = 2.25$$

The speed of P when the string becomes slack is 2.25 m s^{-1}.

Example 9

A particle of mass 0.5 kg is attached to one end A of a light elastic string of natural length 1 m and modulus 5 N. The other end B of the string is fixed to a point on a ceiling. The particle is held at B and then released. Calculate

(a) the length of the string when the particle reaches its lowest point
(b) the speed of the particle when it passes through the equilibrium position.

(a)

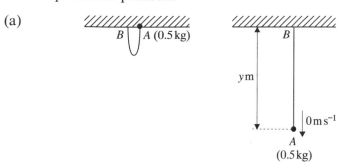

Let the length of the string when the particle is at its lowest point be y m. The extension will then be $(y - 1)$ m. At this lowest point the particle comes instantaneously to rest, that is its speed at its lowest point is $0\,\text{m s}^{-1}$.

Using:
$$\text{E.P.E.} = \frac{\lambda x^2}{2l}$$

with $\lambda = 5\,\text{N}$, $l = 1\,\text{m}$ and $x = (y - 1)\,\text{m}$ gives:

$$\text{E.P.E.} = \frac{5(y - 1)^2}{2 \times 1}\ \text{J}$$

The gravitational P.E. lost by the particle $= mgh$

$$= 0.5 \times 9.8 \times y\ \text{J}$$

Initial K.E. of particle $=$ final K.E. of particle $= 0\,\text{J}$

Hence by the work–energy principle:

E.P.E. gained by string $=$ gravitational P.E. lost by particle

$$\frac{5(y - 1)^2}{2 \times 1} = 0.5 \times 9.8 \times y$$
$$5(y - 1)^2 = 9.8 \times y$$
$$5(y^2 - 2y + 1) - 9.8y = 0$$
$$5y^2 - 19.8y + 5 = 0$$
$$y = \frac{(19.8 \pm \sqrt{(19.8^2 - 4 \times 5 \times 5)})}{2 \times 5}$$
$$y = 0.27 \text{ or } 3.69$$

But y must be greater than 1 m, the natural length of the string.

Hence the length of the string when the particle reaches its lowest point is 3.69 m.

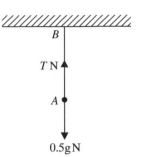

(b) First the equilibrium position must be found.
When the particle is hanging in equilibrium, resolving vertically gives:

$$T = 0.5\,\mathrm{g}$$

Using Hooke's law:
$$T = \frac{\lambda x}{l}$$

with $\lambda = 5\,\mathrm{N}$ and $l = 1\,\mathrm{m}$ gives:

$$T = \frac{5x}{1}$$

So:
$$0.5\mathrm{g} = \frac{5x}{1}$$

$$x = \frac{0.5 \times 9.8}{5} = 0.98$$

The extension in the equilibrium position is 0.98 m and so the length AB is 1.98 m.

Consider the particle falling from rest to its equilibrium position. Let the speed of the particle when AB is 1.98 m be $v\,\mathrm{m\,s^{-1}}$.

Gravitational P.E. lost by particle $= mgh = 0.5 \times 9.8 \times 1.98$ J

K.E. gained by particle $= \frac{1}{2}mv^2 = \frac{1}{2} \times 0.5 \times v^2$ J

E.P.E. of string at the equilibrium position $= \dfrac{\lambda x^2}{2l} = \dfrac{5 \times 0.98^2}{2 \times 1}$ J

By the work–energy principle:

$$(\text{K.E.} + \text{E.P.E.}) \text{ gained} = \text{P.E. lost}$$

$$\frac{1}{2} \times 0.5 \times v^2 + \frac{5 \times 0.98^2}{2 \times 1} = 0.5 \times 9.8 \times 1.98$$

$$v^2 + 10 \times 0.98^2 = 2 \times 9.8 \times 1.98$$

$$v^2 = 2 \times 9.8 \times 1.98 - 10 \times 0.98^2$$

$$v = 5.40$$

The speed of the particle as it passes through its equilibrium position is $5.40\,\mathrm{m\,s^{-1}}$.

Example 10

A particle P of mass 0.8 kg is attached to one end of a light elastic string of natural length 1 m and modulus 20 N. The other end of the string is fixed to a point A on a rough plane inclined at 30° to the horizontal. P is held at rest at A and then released. P comes to instantaneous rest at a point B 1.5 m from A. Calculate the coefficient of friction between P and the plane.

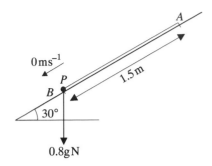

While the particle is moving down the plane there is a frictional force acting up the plane, opposing the motion.

Let this frictional force be F N and the normal reaction between P and the plane be R N.

There will be a tension in the string for part of the motion but this will not affect the calculation of the magnitude of the frictional force.

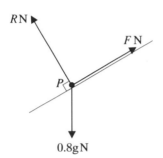

Resolving perpendicular to the plane:

$$R = 0.8 \text{g} \cos 30°$$

Using:

$$F = \mu R$$

where μ is the coefficient of friction gives:

$$F = \mu \times 0.8 \text{g} \cos 30°$$

As the particle travels 1.5 m down the plane, using:

$$\text{work done} = \text{force} \times \text{distance travelled}$$

gives:

work done against frictional force $= \mu \times 0.8 \times 9.8 \cos 30° \times 1.5$ J

The particle starts from rest at A and is at rest again at B. Hence:

$$\text{initial kinetic energy} = \text{final kinetic energy} = 0$$

As the particle moves down the slope from A to B there is a loss of gravitational potential energy. This is given by:

gravitational potential energy lost $= mgh$
$$= 0.8 \times 9.8 \times 1.5 \sin 30° \text{ J}$$

At B, the extension of the string is $(1.5 - 1)\,\text{m} = 0.5\,\text{m}$.

Using:

$$\text{E.P.E.} = \frac{\lambda x^2}{2l}$$

with $\lambda = 20\,\text{N}$, $x = 0.5\,\text{m}$, $l = 1\,\text{m}$ gives:

$$\text{E.P.E.} = \frac{20 \times 0.5^2}{2 \times 1} = 2.5\,\text{J}$$

By the work–energy principle:

work done against frictional force $=$ loss of energy of particle

$$= \text{loss of P.E.} - \text{gain of E.P.E.}$$

Hence:

$$\mu \times 0.8 \times 9.8\cos 30° \times 1.5 = 0.8 \times 9.8 \times 1.5\sin 30° - 2.5$$
$$\mu = \frac{0.8 \times 9.8 \times 1.5\sin 30° - 2.5}{0.8 \times 9.8\cos 30° \times 1.5}$$
$$\mu = 0.3318$$

So the coefficient of friction between P and the plane is 0.332.

Exercise 4D

Whenever a numerical value of g is required, take g $= 9.8\,\text{m}\,\text{s}^{-2}$.

1 An elastic spring of natural length 1 m and modulus 60 N rests on a smooth horizontal floor with one end fixed. A particle of mass 2 kg is attached to the free end of the spring which is compressed to a length of 0.75 m. The particle is released. Calculate the speed of the particle when the spring returns to its natural length.

2 A particle P of mass 3 kg is attached to one end of a light elastic string of natural length 1.5 m and modulus 6 N. The other end of the string is fixed to a point O on a smooth horizontal table. P is released from rest at a point on the table 3 m from O. Calculate the speed of P (a) when the string becomes slack (b) when $OP = 2\,\text{m}$.

3 A particle of mass 0.5 kg is attached to one end A of a light elastic string of natural length 2 m and modulus 20 N. The other end B of the string is fixed to a point on the ceiling. The particle is held at a distance of 1.5 m vertically below B and then released. Calculate (a) the length of the string when the particle reaches its lowest point (b) the speed of the particle when it passes through its equilibrium position.

4   A jack-in-the-box is made using a spring of natural length 0.2 m and modulus 100 N and a 'jack' of mass 0.5 kg. When the lid is closed the spring is compressed to a length of 0.1 m. Assuming the spring to be vertical throughout, calculate the maximum distance that the 'jack' will rise when the lid is suddenly raised.

5   A particle P of mass m is attached to one end of a light elastic string of natural length l whose other end is attached to a point A on a ceiling. When P hangs in equilibrium AP has length $\frac{5l}{3}$. Show that if P is projected vertically downwards from A with speed $\sqrt{\left(\frac{3gl}{2}\right)}$, P will first come to instantaneous rest after moving a distance $\frac{10l}{3}$.

6   A particle of mass m is attached to one end of an elastic spring of natural length l and modulus λ. The particle and spring rest on a rough horizontal surface and the other end of the spring is fixed. The coefficient of friction between the particle and the surface is μ. The particle is held at rest with the spring compressed to a length $\frac{2}{3}l$ and then released. Show that the particle is instantaneously at rest when it has moved a distance x, where

$$x = \frac{2\lambda l - 6\mu lmg}{3\lambda}$$

7   A particle P of mass m is attached to one end of a light elastic string of natural length l and modulus λ. The other end of the string is fixed to a point A on a smooth plane inclined at $\arcsin\frac{3}{5}$ to the horizontal. P is held at rest on the inclined plane with AP along a line of greatest slope of the plane. P is released and first comes to instantaneous rest after moving a distance x.
(a) When P is released from a point below A such that $AP = l$ show that $x = \frac{6mgl}{5\lambda}$.
(b) Given that $\lambda = \frac{6mg}{5}$, find x in terms of l when P is released from a point above A such that $AP = l$.

8   A particle P of mass 2 kg is attached to one end of a light elastic string of natural length 0.6 m and modulus 20 N. One end of the string is fixed to a point A on a rough plane inclined at 45° to

the horizontal. P is held at rest on the plane below A with AP along a line of greatest slope of the plane and $AP = 0.6\,\text{m}$. The coefficient of friction between P and the plane is 0.2. P is released. Calculate the distance P moves down the plane before coming to instantaneous rest. Calculate also the speed of P when $AP = 0.8\,\text{m}$.

9 A particle P of mass 0.3 kg is suspended by two identical elastic strings of natural length 1.5 m and modulus $\lambda\,\text{N}$. The other ends of the strings are fixed to two points A and B on a horizontal ceiling where $AB = 3\,\text{m}$. P is released from rest at the mid-point of AB and falls vertically until it is instantaneously at rest at a point 1 m below the level of the ceiling. Calculate the value of λ and the speed of P when it is 0.5 m below the level of the ceiling.

10 A boy makes a catapult by attaching the ends of a piece of elastic of natural length 0.3 m and modulus 60 N to the ends of a forked stick which are 0.2 m apart. He takes a stone of mass 0.1 kg and places it at the mid-point of the elastic, before pulling the stone backwards until the elastic measures 0.75 m, keeping the elastic horizontal. The stone is then released.

(a) Briefly describe the mathematical model you would use to investigate the motion of the stone, stating any assumptions you would make about the stone or the elastic.

(b) Calculate the speed of the stone on release.

(c) The boy keeps the elastic horizontal. Would the answer to (b) be different if the elastic were vertical? Briefly justify your answer.

11 A light elastic string has natural length l and modulus λ. One end is fixed to a point A on a ceiling and a particle P of mass m is attached to the other end. P is held vertically below A so that $AP = 2l$ and then released. P has speed v when the extension of the string is x. Show that, while the string remains taut,

$$\frac{1}{2}mv^2 = \frac{1}{2}l(\lambda - 2mg) + mgx - \frac{1}{2}\lambda\frac{x^2}{l}$$

By considering the speed of P when $x = 0$ show that the string will never become slack provided $2mg > \lambda$.

SUMMARY OF KEY POINTS

Motion of a particle in a vertical circle

1   A particle which is moving in a *vertical* circle has variable speed.

2   The acceleration of a particle moving in a circle of radius r with speed v at time t has two components:

$$\frac{v^2}{r} \quad \text{towards the centre of the circle}$$

and: $\quad \dfrac{dv}{dt}$ along the tangent.

3   The component of the force along the radius does no work during the motion.

4   By the work–energy principle, the sum of the K.E. and P.E. of the particle is constant throughout the motion.

5   A particle which cannot leave its vertical circular path (for example a particle on a rod) will describe complete circles provided its velocity at the highest point of the circle is greater than or equal to zero.

6   A particle which can leave its vertical circular path (for example a particle on a string) will do so when the force towards the centre of the circle becomes zero.

Energy stored in an elastic string or spring

7   The work done in stretching (or compressing) an elastic string (or spring) with modulus λ from its natural length l to a length $(l+x)$ (or $(l-x)$) is $\dfrac{\lambda x^2}{2l}$.

8   The elastic energy (E.P.E.) in a string (or spring) of modulus λ extended (or compressed) by a length x beyond its natural length is also $\dfrac{\lambda x^2}{2l}$.

9   The total change of the mechanical energies (that is kinetic, gravitational potential and elastic potential energies) of a system is equal to the work done by any external forces acting on the system.

Collisions

5

5.1 Collisions in the real world

Collisions between two freely moving bodies or between a freely moving body and a fixed object are everyday occurrences. In general as a result of a collision the bodies bounce away from each other. Consider the collision of two tennis balls. The time during which they are in contact may be divided into two parts:

(1) the period of compression,

(2) the period of restitution, during which they recover their shape.

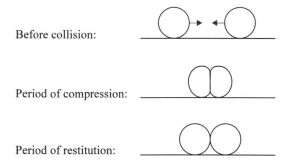

Before collision:

Period of compression:

Period of restitution:

The property which causes bodies like tennis balls to recover their shape and hence causes them to rebound is called **elasticity**.

In dealing with the impact of elastic bodies it is usually assumed that they are **smooth**. This means that the mutual reaction acts along the common normal at the point of impact. When two elastic spheres collide the mutual reaction acts along their line of centres as this is the common normal. The bodies are said to collide **directly** if the motion of each is along this common normal. For example two spheres of the same radius collide directly if they are moving parallel to their line of centres.

In this chapter we will consider the collision of 'elastic particles' moving on a straight line. Many collisions in the real world may be modelled by this situation by making appropriate assumptions, for example the collision of two snooker balls.

5.2 Collisions of particles

In Chapter 5 of Book M1 the collision of two elastic particles was considered. Remember that:

(1) The momentum of a particle of mass m moving with velocity $\mathbf{v}$ is $m\mathbf{v}$. As $\mathbf{v}$ is a vector it follows that **momentum is also a vector**.

(2) **The impulse of a force = change in momentum produced**

$$\mathbf{I} = m\mathbf{v} - m\mathbf{u}$$

This is sometimes called the **impulse–momentum principle**.

(3) The **principle of conservation of momentum**:

■ **Total momentum before collision = total momentum after collision**

For particles moving along a straight line this gives:

Before collision:

$\longrightarrow u_1 \qquad \longrightarrow u_2$

$\bigcirc \qquad\qquad \bigcirc$

$m_1 \qquad\qquad m_2$

After collision:

$\longrightarrow v_1 \qquad \longrightarrow v_2$

$\bigcirc \qquad\qquad \bigcirc$

$m_1 \qquad\qquad m_2$

$$m_1 u_1 + m_2 u_2 = m_1 v_1 + m_2 v_2$$

Example 1

A particle of mass 5 kg is moving with velocity $(3\mathbf{i} + 4\mathbf{j})\,\mathrm{m\,s^{-1}}$ when it is given an impulse $(-2\mathbf{i} + 6\mathbf{j})\,\mathrm{N\,s}$. Find the velocity of the particle after the impact.

If the velocity of the particle after impact is $\mathbf{v}\,\mathrm{m\,s^{-1}}$ then the change in momentum is:

$$5[\mathbf{v} - (3\mathbf{i} + 4\mathbf{j})]\,\mathrm{kg\,m\,s^{-1}}$$

From the impulse–momentum principle this is equal to the impulse.

So:
$$5\mathbf{v} - 5(3\mathbf{i} + 4\mathbf{j}) = -2\mathbf{i} + 6\mathbf{j}$$
$$5\mathbf{v} = -2\mathbf{i} + 6\mathbf{j} + 15\mathbf{i} + 20\mathbf{j}$$
$$= 13\mathbf{i} + 26\mathbf{j}$$

and:
$$\mathbf{v} = \frac{1}{5}(13\mathbf{i} + 26\mathbf{j})$$

The velocity of the particle after impact is $\frac{1}{5}(13\mathbf{i} + 26\mathbf{j})\,\mathrm{m\,s^{-1}}$.

Example 2

A particle of mass 0.5 kg is moving horizontally in a straight line with a speed of $20\,\mathrm{m\,s^{-1}}$. It is hit by a bat, and moves back along its original path with a speed of $25\,\mathrm{m\,s^{-1}}$. Find the magnitude of the impulse exerted on the particle by the bat. Describe a 'realistic situation' this could be used to model.

The motion in this problem is along a straight line. The vector properties of the momentum and the impulse may be taken into account by taking the direction of the impulse as positive.

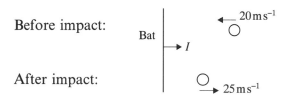

Therefore:

change in momentum is $= 0.5(25 - (-20))$

$$= 22.5$$

This is equal to the impulse, so:

$$\mathbf{I} = 22.5\,\mathrm{N\,s}$$

The magnitude of the impulse exerted by the bat on the particle is $22.5\,\mathrm{N\,s}$.

This could be used to model a cricket bat striking a cricket ball. The ball will then be modelled by a particle. This is a reasonable assumption as the dimensions of the ball are small compared to those of the bat.

Example 3

Two particles P and Q of mass 2 kg and 3 kg respectively are moving towards each other along the same straight line with speeds $4\,\mathrm{m\,s^{-1}}$ and $5\,\mathrm{m\,s^{-1}}$ respectively. After the collision the particles coalesce, that is move as a single body. Find the speed of the combined particle after the collision. Describe a situation this could be used to model.

It is important to remember that momentum is a vector. The equation:

total momentum before collision $=$ total momentum after collision

is therefore a **vector equation**.

As in the previous example it is necessary to choose a particular direction as positive. It is immaterial which direction is chosen.

Choose the direction of motion of P as positive and let the speed of the combined particle after impact be $v\,\mathrm{m\,s^{-1}}$ in that direction. The figure shows the situation before and after impact.

Before impact:

$\overset{4\,\mathrm{m\,s^{-1}}}{\longrightarrow}$ $\overset{5\,\mathrm{m\,s^{-1}}}{\longleftarrow}$

$\bigcirc$ $\bigcirc$

$P\,(2\,\mathrm{kg})$ $Q\,(3\,\mathrm{kg})$

After impact:

$\longrightarrow v\,\mathrm{m\,s^{-1}}$

$\bigcirc$

$P{+}Q\,(5\,\mathrm{kg})$

The momentum before impact $= 2 \times 4 - 3 \times 5$

The momentum after impact $= 5v$

By the principle of conservation of momentum:

$$2 \times 4 - 3 \times 5 = 5v$$

So:
$$v = \frac{-7}{5} = -1.4$$

The speed of the combined particle is $1.4\,\mathrm{m\,s^{-1}}$. Since v is negative, the speed is in the opposite direction to the original motion of particle P.

This could be used to model two balls of putty which will stick together on impact.

It could also be used to model two small trucks, moving along a track, which couple when they collide.

Exercise 5A

1 A particle of mass $0.5\,\mathrm{kg}$ is moving with velocity $(6\mathbf{i} + 2\mathbf{j})\,\mathrm{m\,s^{-1}}$ when it is given an impulse $(3\mathbf{i} - 2\mathbf{j})\,\mathrm{N\,s}$. Find the new velocity of the particle. Describe a 'realistic situation' this could model.

2 A particle of mass $0.5\,\mathrm{kg}$ is moving with velocity $(4\mathbf{i} + 7\mathbf{j})\,\mathrm{m\,s^{-1}}$ when it strikes a fixed wall. It rebounds with velocity $(2\mathbf{i} + 3\mathbf{j})\,\mathrm{m\,s^{-1}}$. Find the impulse exerted on the particle by the wall.

3 An impulse $(4\mathbf{i} - 8\mathbf{j})\,\mathrm{N\,s}$ is applied to a particle of mass $2\,\mathrm{kg}$. The velocity of the mass after the impact is $(3\mathbf{i} + 5\mathbf{j})\,\mathrm{m\,s^{-1}}$. Find its velocity before impact.

4 A particle of mass $1.5\,\mathrm{kg}$ is moving along a straight line with speed $12\,\mathrm{m\,s^{-1}}$. It receives an impulse $I\,\mathrm{N\,s}$ and continues to move in the same direction but with a speed of $24\,\mathrm{m\,s^{-1}}$. Find I.

5 A particle of mass 3 kg is moving along a straight line with speed $8\,\text{m s}^{-1}$. It receives an impulse I N s and then moves along the same straight line but in the opposite direction, with a speed of $4\,\text{m s}^{-1}$. Find I.

6 A particle of mass 2 kg is falling at a speed of $12\,\text{m s}^{-1}$ when it strikes the ground. The particle is brought to rest by the impact. Find the impulse exerted by the ground on the particle. Describe a 'realistic situation' this could model.

7 A particle of mass 3 kg falls from rest at a height of 5 m above a horizontal plane. It rebounds to a height of 3 m. Find the impulse exerted on the particle by the plane.

8 Two particles A and B of mass 3 kg and 4 kg respectively are moving towards each other along the same straight line with speeds $5\,\text{m s}^{-1}$ and $3\,\text{m s}^{-1}$ respectively. After the collision the particles coalesce. Find the magnitude and direction of the velocity of the combined particle after the impact. What realistic situation could this model?

9 Two particles P and Q of mass 4 kg and 2 kg respectively travel towards each other along the x-axis. The velocities of P and Q are $6\mathbf{i}\,\text{m s}^{-1}$ and $(-2\mathbf{i})\,\text{m s}^{-1}$ respectively. The particles collide and after the collision Q has a velocity $3\mathbf{i}\,\text{m s}^{-1}$. Find the velocity of P after the collision.

10 Two particles A and B of mass 3 kg and 5 kg respectively move along the same straight line and collide. Just before the collision A is moving with speed $8\,\text{m s}^{-1}$ and B is moving with speed $4\,\text{m s}^{-1}$ in the opposite direction. After the collision the particle A rebounds with speed $7\,\text{m s}^{-1}$. Find the velocity of B after the collision.

5.3 Newton's law of restitution for direct impact

A **direct** impact is a collision between two particles which are moving along the same straight line.

When a ball strikes a fixed wall the speed with which it rebounds depends on the material of which the ball is made. In a similar way when two elastic particles collide their speeds after the collision depend on the material they are made from.

Newton's law of restitution, sometimes called Newton's experimental law, defines precisely how the speeds of the particles after the collision depend on the nature of the particles as well as on the speeds before the collision. It states that:

■ $$\frac{\text{speed of separation of particles}}{\text{speed of approach of particles}} = e$$

Notice that both the numerator and the denominator are positive in the statement.

The constant e is called the **coefficient of restitution**.

This law only holds when the collision takes place in free space or on a **smooth** surface. (On a **rough** surface the speed of separation of two particles may be affected by 'spin'.) The value of e satisfies the inequality

$$0 \leqslant e \leqslant 1$$

For two perfectly elastic particles $e = 1$.

For two inelastic particles, that is two particles which coalesce on impact, $e = 0$.

For a pair of glass marbles $e = 0.95$.

For a pair of billiard balls $e = 0.8$.

For a pair of lead spheres $e = 0.2$.

It is important to be clear what 'the speed of separation' and 'the speed of approach' are in any given situation. The diagrams below illustrate three typical situations.

(1) Before impact:

$\longrightarrow u_1 \qquad \longrightarrow 0 \text{ (at rest)}$

Speed of approach $= u_1 - 0 = u_1$

After impact:

$\longrightarrow 0 \text{ (at rest)} \qquad \longrightarrow v_2$

Speed of separation $= v_2 - 0 = v_2$

(2) Before impact:

$\longrightarrow u_1 \qquad \longrightarrow u_2$

Speed of approach $= u_1 - u_2$

(u_1 must be greater than u_2 or there would be no collision.)

After impact: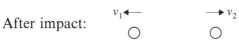

Speed of separation $= v_2 - v_1$

(v_2 must be greater than v_1 or they would not separate.)

(3) Before impact:

Speed of approach $= u_1 + u_2$

After impact:

Speed of separation $= v_1 + v_2$

Example 4

Find the value of e in the situations shown.

(a) Before impact:
→ 6 m s⁻¹   → 0 (at rest)

After impact:
→ 2 m s⁻¹   → 4 m s⁻¹

(b) Before impact:
→ 4 m s⁻¹   2 m s⁻¹ ←

After impact:
1 m s⁻¹ ←   → 3 m s⁻¹

(a)

$$e = \frac{\text{speed of separation}}{\text{speed of approach}}$$

$$= \frac{4 - 2}{6 - 0} = \frac{2}{6} = \frac{1}{3}$$

(b)

$$e = \frac{\text{speed of separation}}{\text{speed of approach}}$$

$$= \frac{3 + 1}{4 + 2} = \frac{4}{6} = \frac{2}{3}$$

Example 5

Find the value of v in the situation shown, given that the coefficient of restitution e is $\frac{2}{3}$.

Before impact: $\longrightarrow 6\,\text{m s}^{-1}$ $3\,\text{m s}^{-1}\longleftarrow$
 ○   ○

After impact: $\longrightarrow v\,\text{m s}^{-1}$ $\longrightarrow 7\,\text{m s}^{-1}$
 ○   ○

As:
$$e = \frac{\text{speed of separation}}{\text{speed of approach}}$$

so:
$$\frac{2}{3} = \frac{7-v}{6+3}$$

and:
$$2(9) = 3(7-v)$$

Therefore:
$$v = 1$$

Problems involving the direct collision of two particles may be solved by writing down two equations:

(1) The conservation of linear momentum,
(2) Newton's law of restitution.

These two simultaneous equations may be solved to find two unknowns.

Example 6

Two particles A and B of mass $0.2\,\text{kg}$ and $0.5\,\text{kg}$ respectively are moving towards each other along the same straight line on a smooth horizontal table. Particle A has speed $12\,\text{m s}^{-1}$ and particle B has speed $2\,\text{m s}^{-1}$. Given that the coefficient of restitution between the particles is $\frac{1}{2}$, find:

(a) the speeds of A and B after the impact,
(b) the magnitude of the impulse given to each particle.

Describe a realistic situation this could be used to model.

(a) First draw diagrams of the situations before and after impact.

Before impact: $\longrightarrow 12\,\text{m s}^{-1}$ $2\,\text{m s}^{-1}\longleftarrow$
 ○   ○
 $A\,(0.2\,\text{kg})$ $B\,(0.5\,\text{kg})$

After impact: $\longrightarrow v_1\,\text{m s}^{-1}$ $v_2\,\text{m s}^{-1}\longleftarrow$
 ○   ○
 $A\,(0.2\,\text{kg})$ $B\,(0.5\,\text{kg})$

Using conservation of momentum with $m_1 = 0.2\,\text{kg}$, $m_2 = 0.5\,\text{kg}$, $u_1 = 12\,\text{m}\,\text{s}^{-1}$ and $u_2 = -2\,\text{m}\,\text{s}^{-1}$ gives:

$$(0.2)12 - (0.5)2 = (0.2)v_1 + (0.5)v_2$$

or: $\qquad\qquad 2v_1 + 5v_2 = 14 \qquad\qquad\qquad (1)$

Newton's law of restitution states:

$$e = \frac{\text{speed of separation}}{\text{speed of approach}}$$

So: $\qquad\qquad \dfrac{1}{2} = \dfrac{v_2 - v_1}{14}$

and: $\qquad\qquad v_2 - v_1 = 7 \qquad\qquad\qquad\qquad (2)$

Eliminating v_1 between equations (1) and (2) gives:

$$7v_2 = 28$$

So: $\qquad\qquad v_2 = 4$

Substituting this value into equation (2) gives:

$$v_1 = -3$$

The minus sign indicates that A moves in the opposite direction to the arrow in the diagram. The situation after collision is:

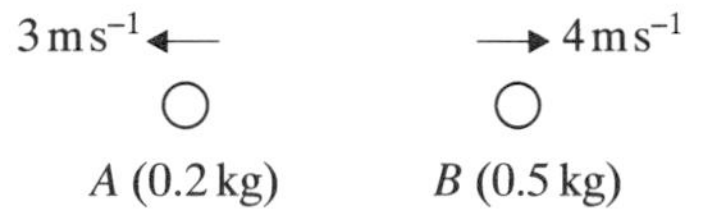

$3\,\text{m}\,\text{s}^{-1} \leftarrow \qquad\qquad \rightarrow 4\,\text{m}\,\text{s}^{-1}$

$\bigcirc \qquad\qquad \bigcirc$

$A\ (0.2\,\text{kg}) \qquad B\ (0.5\,\text{kg})$

In this case the direction of motion of each particle is reversed by the collision.

(b) Recall that 'Impulse = change in momentum' and also that 'momentum is a vector'. Take the initial direction of the velocity of A as positive, that is from left to right.

For particle A we have:

$\rightarrow 12\,\text{m}\,\text{s}^{-1}$

Before impact: $\qquad \bigcirc$

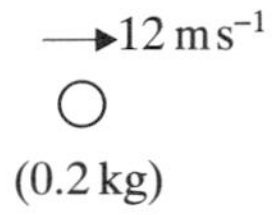

$(0.2\,\text{kg})$

$3\,\text{m}\,\text{s}^{-1} \leftarrow$

After impact: $\qquad \bigcirc$

$(0.2\,\text{kg})$

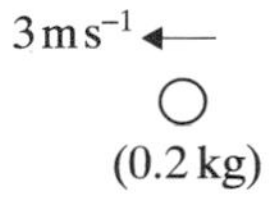

The change in momentum of A is:

$$-(0.2)3 - (0.2)(12) = -3\,\text{N}\,\text{s}$$

So magnitude of impulse on $A = 3\,\mathrm{N\,s}$

For particle B we have:

Before impact:

$2\,\mathrm{ms}^{-1}\longleftarrow$

$\bigcirc$

$(0.5\,\mathrm{kg})$

After impact:

$\longrightarrow 4\,\mathrm{ms}^{-1}$

$\bigcirc$

$(0.5\,\mathrm{kg})$

The change in momentum of B is:

$$(0.5)4 - (0.5)(-2) = 3\,\mathrm{N\,s}$$

So magnitude of impulse on $B = 3\,\mathrm{N\,s}$.

The two particles A and B may be used to model two small smooth spheres, with the same radius, of mass $0.2\,\mathrm{kg}$ and $0.5\,\mathrm{kg}$ respectively, moving towards each other on a sheet of ice.

Alternatively A and B may be used to model two toy trucks of mass $0.2\,\mathrm{kg}$ and $0.5\,\mathrm{kg}$ respectively moving towards each other on a smooth horizontal straight track.

Loss of mechanical energy due to impact

When two elastic particles collide although there is no change in momentum there is in general, $(e \neq 1)$, a loss of mechanical energy. When $e = 1$ the particles are perfectly elastic and there is no loss of mechanical energy. (See question 8 in Exercise 5B.) In reality $e \neq 1$ and some of the kinetic energy of the particles is transformed into other forms of energy such as heat energy or sound energy.

For particles moving along a straight line:

Before impact:

$\longrightarrow u_1$ $\longrightarrow u_2$

$\bigcirc$ $\bigcirc$

m_1 m_2

After impact:

$\longrightarrow v_1$ $\longrightarrow v_2$

$\bigcirc$ $\bigcirc$

m_1 m_2

The total kinetic energy before impact is:

$$\tfrac{1}{2}m_1 u_1^2 + \tfrac{1}{2}m_2 u_2^2$$

The total kinetic energy after impact is:

$$\tfrac{1}{2}m_1 v_1^2 + \tfrac{1}{2}m_2 v_2^2$$

The loss of mechanical energy due to impact is therefore:

$$\left[\tfrac{1}{2}m_1u_1^2 + \tfrac{1}{2}m_2u_2^2\right] - \left[\tfrac{1}{2}m_1v_1^2 + \tfrac{1}{2}m_2v_2^2\right]$$

For the particles in Example 6 this loss is:

$$\left[\tfrac{1}{2}(0.2)(12)^2 + \tfrac{1}{2}(0.5)(2)^2\right] - \left[\tfrac{1}{2}(0.2)(3)^2 + \tfrac{1}{2}(0.5)(4)^2\right] = 10.5\,\text{J}$$

The change in mechanical energy may also be obtained from:

[change in kinetic energy of particle A] + [change in kinetic energy of particle B]

For the particles in Example 6 this gives:

$$\tfrac{1}{2}(0.2)\left[(12)^2 - (3)^2\right] + \tfrac{1}{2}(0.5)\left[(2)^2 - (4)^2\right] = 10.5\,\text{J}$$

Exercise 5B

1 In each part of this question the two diagrams show the speeds of two particles A and B just before and just after a collision. The particles move on a smooth horizontal plane. Find the coefficient of restitution e in each case.

Before collision: After collision:

(a)
$\longrightarrow 12\,\text{m s}^{-1}$ At rest   At rest $\longrightarrow 6\,\text{m s}^{-1}$
A B A B

(b)
$\longrightarrow 6\,\text{m s}^{-1}$ $\longrightarrow 2\,\text{m s}^{-1}$ $\longrightarrow 3\,\text{m s}^{-1}$ $\longrightarrow 4\,\text{m s}^{-1}$
A B A B

(c)
$\longrightarrow 5\,\text{m s}^{-1}$ $\longleftarrow 4\,\text{m s}^{-1}$ $3\,\text{m s}^{-1}\longleftarrow$ $\longrightarrow 3\,\text{m s}^{-1}$
A B A B

2 In each part of this question the two diagrams show the speeds of two particles A and B just before and just after collision on a smooth horizontal plane. The masses of A and B and the coefficient of restitution e are also given. Find the speeds v_1 and v_2 in each case.

Before collision: After collision:

(a) $e = \frac{1}{2}$: $\xrightarrow{\ 8\,\mathrm{m\,s^{-1}}}$ O   At rest O $\xrightarrow{\ v_1\,\mathrm{m\,s^{-1}}}$ O $\xrightarrow{\ v_2\,\mathrm{m\,s^{-1}}}$ O
$A\,(0.5\,\mathrm{kg})$ $B\,(1\,\mathrm{kg})$ $A\,(0.5\,\mathrm{kg})$ $B\,(1\,\mathrm{kg})$

(b) $e = \frac{1}{4}$: $\xrightarrow{\ 6\,\mathrm{m\,s^{-1}}}$ O $\xrightarrow{\ 2\,\mathrm{m\,s^{-1}}}$ O $\xrightarrow{\ v_1\,\mathrm{m\,s^{-1}}}$ O $\xrightarrow{\ v_2\,\mathrm{m\,s^{-1}}}$ O
$A\,(2\,\mathrm{kg})$ $B\,(3\,\mathrm{kg})$ $A\,(2\,\mathrm{kg})$ $B\,(3\,\mathrm{kg})$

(c) $e = \frac{1}{2}$: $\xrightarrow{\ 4\,\mathrm{m\,s^{-1}}}$ O $\xleftarrow{\ 1\,\mathrm{m\,s^{-1}}}$ O $\xrightarrow{\ v_1\,\mathrm{m\,s^{-1}}}$ O $\xrightarrow{\ v_2\,\mathrm{m\,s^{-1}}}$ O
$A\,(4\,\mathrm{kg})$ $B\,(1\,\mathrm{kg})$ $A\,(4\,\mathrm{kg})$ $B\,(1\,\mathrm{kg})$

(d) $e = \frac{3}{4}$: $\xrightarrow{\ 3\,\mathrm{m\,s^{-1}}}$ O $\xleftarrow{\ 1\,\mathrm{m\,s^{-1}}}$ O $\xrightarrow{\ v_1\,\mathrm{m\,s^{-1}}}$ O $\xrightarrow{\ v_2\,\mathrm{m\,s^{-1}}}$ O
$A\,(1\,\mathrm{kg})$ $B\,(2.5\,\mathrm{kg})$ $A\,(1\,\mathrm{kg})$ $B\,(2.5\,\mathrm{kg})$

(e) $e = \frac{1}{5}$: $\xrightarrow{\ 2\,\mathrm{m\,s^{-1}}}$ O $\xleftarrow{\ 3\,\mathrm{m\,s^{-1}}}$ O $\xrightarrow{\ v_1\,\mathrm{m\,s^{-1}}}$ O $\xrightarrow{\ v_2\,\mathrm{m\,s^{-1}}}$ O
$A\,(2\,\mathrm{kg})$ $B\,(2\,\mathrm{kg})$ $A\,(2\,\mathrm{kg})$ $B\,(2\,\mathrm{kg})$

3 A small smooth sphere A of mass $2\,\mathrm{kg}$ is travelling along a straight line on a smooth horizontal plane with speed $6\,\mathrm{m\,s^{-1}}$ when it collides with a small smooth sphere B of mass $3\,\mathrm{kg}$ moving along the same straight line in the same direction with speed $2\,\mathrm{m\,s^{-1}}$. After the collision A continues to move in the same direction with speed $3\,\mathrm{m\,s^{-1}}$.
(a) Find the speed of B after the collision.
(b) Find the coefficient of restitution between A and B.
(c) State any assumptions you made in your calculations.

4 Two smooth spheres of equal radius have mass $2.5\,\mathrm{kg}$ and $1.5\,\mathrm{kg}$ respectively. They are travelling towards each other along the same straight line on a smooth horizontal plane. The heavier sphere has a speed of $3\,\mathrm{m\,s^{-1}}$ and the other sphere has a speed of $2\,\mathrm{m\,s^{-1}}$. The spheres collide and after the collision the heavier sphere is at rest.
(a) Find the coefficient of restitution between the spheres.
(b) Find the kinetic energy lost in the collision.

5 A small smooth sphere of mass $3\,\mathrm{kg}$ moving on a smooth horizontal plane with a speed of $8\,\mathrm{m\,s^{-1}}$ collides directly with a sphere of mass $12\,\mathrm{kg}$ which is at rest. Given that the spheres

move in opposite directions after the collision, obtain the inequality satisfied by e.

6 Two identical smooth spheres each of mass m are projected directly towards each other on a smooth horizontal surface. Each sphere has a speed u and the coefficient of restitution between the spheres is e. Show that the collision between the spheres causes a loss of kinetic energy of $mu^2(1 - e^2)$. Describe a 'realistic situation' this could be used to model. State clearly the assumptions you have made.

7 A sphere of mass m is moving with a speed V along a horizontal straight line. It collides with an identical sphere of mass m moving along the same straight line in the same direction with speed $u\,(u < V)$. Show that the magnitude of the impulse on either of the spheres is

$$\frac{1}{2}m(1 + e)(V - u)$$

where e is the coefficient of restitution between the two spheres.

8 A sphere of mass m_1 moving with speed u_1 collides directly with a similar sphere of mass m_2 moving with speed u_2 in the same direction $(u_1 > u_2)$. The coefficient of restitution between the two spheres is e. Show that the loss of kinetic energy E due to the collision satisfies the equation

$$2(m_1 + m_2)E = m_1 m_2 (u_1 - u_2)^2 (1 - e^2)$$

5.4 Impact of a particle with a fixed surface

Consider a small sphere, moving on a smooth horizontal surface, which strikes a fixed vertical barrier that is perpendicular to the direction of motion of the sphere as shown in the figure.

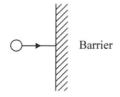
Barrier

As the sphere is small we can model it by an elastic particle. The phrase commonly used to describe the above situation is 'the particle collides normally with a fixed barrier'.

From your own experience you will know that the direction of motion of the sphere is reversed by the impact. The speed of the sphere after the impact will depend on the material the sphere is made from.

Let the speed before impact be u, the speed after impact be v, and the coefficient of restitution between the particle and the surface be e.

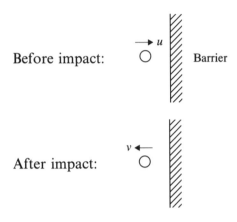

Before impact:

After impact:

Newton's law of restitution applied to this impact gives:

$$e = \frac{\text{speed of separation}}{\text{speed of approach}}$$

$$= \frac{v - 0}{u - 0} = \frac{v}{u}$$

$$= \frac{\text{speed of rebound}}{\text{speed of approach}}$$

Hence

$$e = \frac{v}{u} \quad \text{or} \quad v = eu$$

So: (speed of rebound) $= e \times$ (speed of approach)

Example 7

A particle is moving horizontally with speed $5\,\text{m s}^{-1}$ when it strikes a vertical wall normally.

It rebounds with a speed of $3\,\text{m s}^{-1}$. Find the coefficient of restitution between the particle and the wall.

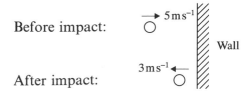

Before impact:

After impact:

By Newton's law of restitution:

$$e = \frac{\text{speed of rebound}}{\text{speed of approach}}$$
$$= \frac{3}{5}$$

The coefficent of restitution is $\frac{3}{5}$.

Example 8

A particle is moving horizontally with speed $3\,\text{m}\,\text{s}^{-1}$ when it strikes a vertical wall normally. The coefficient of restitution between the particle and the wall is $\frac{3}{4}$. Find the speed of rebound.

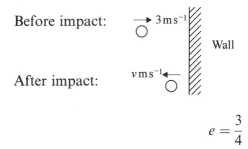

Before impact:

After impact:

$$e = \frac{3}{4}$$

By Newton's law of restitution:

$$e = \frac{\text{speed of rebound}}{\text{speed of approach}}$$

In this case we have:

$$\frac{3}{4} = \frac{v - 0}{3 - 0}$$

So:
$$v = \frac{9}{4}$$

The speed of rebound is $2\frac{1}{4}\,\text{m}\,\text{s}^{-1}$.

Example 9

A small smooth ball falls from rest from a height of $1.35\,\text{m}$ above a fixed smooth horizontal plane. It rebounds to a height of $0.6\,\text{m}$. Find the coefficient of restitution between the ball and the plane.

Model the ball by an 'elastic particle'.

The particle moves freely under gravity until it strikes the plane. The speed $v\,\text{m s}^{-1}$ with which it strikes the plane can be found using the constant acceleration formula (Book M1, Chapter 3).

Using $v^2 = u^2 + 2as$ with $u = 0\,\text{m s}^{-1}$, $s = 1.35\,\text{m}$ and $a = 9.8\,\text{m s}^{-2}$ gives:

$$v^2 = 2(9.8)(1.35) \qquad (1)$$

Suppose the particle rebounds from the plane with speed $u\,\text{m s}^{-1}$. Then:

$$e = \frac{u}{v} \text{ or } u = ev \qquad (2)$$

After rebound the particle again moves freely under gravity. Since at the greatest height the speed of the particle is zero, using $v^2 = u^2 + 2as$ with $v = 0\,\text{m s}^{-1}$, $s = 0.6\,\text{m}$ and $a = -9.8\,\text{m s}^{-2}$ gives:

$$u^2 = 2(9.8)(0.6) \qquad (3)$$

Substituting (2) into (3) gives:

$$e^2 v^2 = 2(9.8)(0.6) \qquad (4)$$

and substituting (1) into (4) gives:

$$e^2 2(9.8)(1.35) = 2(9.8)(0.6)$$

or:

$$e^2 = \frac{0.6}{1.35} = \frac{4}{9}$$

So:

$$e = \frac{2}{3} \qquad \text{since } 0 \leqslant e \leqslant 1$$

The coefficient of restitution between the ball and the plane is $\frac{2}{3}$.

Example 10

A particle P falls from rest from a height of $4\,\text{m}$ above a smooth horizontal plane. The coefficient of restitution between P and the plane is $\frac{1}{2}$. Find the total distance travelled by P up to the instant when it hits the plane for the 3rd time.

1st rebound

The particle moves freely under gravity until it strikes the plane. The speed with which it strikes the plane may be obtained by using the constant acceleration formula (Book M1, Chapter 3) $v^2 = u^2 + 2as$ with $u = 0\,\text{m s}^{-1}$, $s = 4\,\text{m}$ and $a = 9.8\,\text{m s}^{-2}$.

So:
$$v^2 = 2(9.8)4$$

and:
$$v = 2(19.6)^{\frac{1}{2}}\,\mathrm{m\,s}^{-1} \qquad (1)$$

Using Newton's law of restitution the speed of rebound is $v_1 = ev$ where v is given by (1).

So:
$$v_1 = \frac{1}{2}2(19.6)^{\frac{1}{2}}\,\mathrm{m\,s}^{-1} \qquad (2)$$

The height h_1 to which the particle will rise after the first rebound is obtained by using $v^2 = u^2 + 2as$ with $v = 0\,\mathrm{m\,s}^{-1}$, $u = v_1 = (19.6)^{\frac{1}{2}}\,\mathrm{m\,s}^{-1}$ and $a = -9.8\,\mathrm{m\,s}^{-2}$.

So:
$$0 = 19.6 - 2(9.8)h_1$$

and:
$$h_1 = 1\,\mathrm{m} \qquad (3)$$

The particle rises to a height of 1 m after the 1st rebound. The figure summarises the results so far.

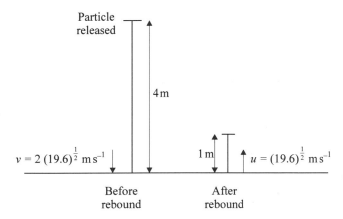

2nd rebound

The particle will now fall freely under gravity and reach the plane for a second time. By symmetry it will reach the plane with a speed of magnitude $v_1 = (19.6)^{\frac{1}{2}}\,\mathrm{m\,s}^{-1}$ as given by (2).

Using Newton's law of restitution the speed of the second rebound v_2 is:
$$v_2 = ev_1$$

So:
$$v_2 = \frac{1}{2}(19.6)^{\frac{1}{2}}\,\mathrm{m\,s}^{-1} \qquad (4)$$

The height h_2 to which the particle will rise after the second rebound is obtained as before by using $v^2 = u^2 + 2as$ but now with $v = 0\,\mathrm{m\,s}^{-1}$, $u = v_2 = \frac{1}{2}(19.6)^{\frac{1}{2}}\,\mathrm{m\,s}^{-1}$ and $a = -9.8\,\mathrm{m\,s}^{-2}$.

So:
$$0 = \frac{19.6}{4} - 2(9.8)h_2$$

and:
$$h_2 = \frac{1}{4}\,\text{m} = 0.25\text{m} \qquad (5)$$

The particle rises to a height of 0.25 m after the second rebound. The figure summarises the results for the second rebound.

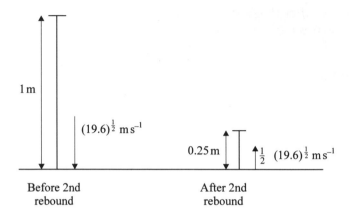

The particle will now fall freely under gravity a distance of 0.25 m and reach the plane for a third time.

The total distance travelled by the particle up to the instant when it hits the plane for the third time is:

(distance travelled before 1st rebound)

+ (distance travelled between 1st and 2nd rebounds)

+ (distance travelled between 2nd and 3rd rebounds)

Using (3) and (5) gives:

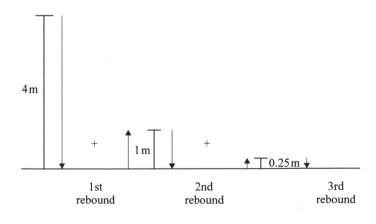

Hence total distance is:

$$4 + (2 \times 1) + (2 \times 0.25) = 4 + 2 + 0.5 = 6.5 \, \text{m}$$

The total distance travelled by the particle up to the instant when it hits the plane for the 3rd time is 6.5 m.

Exercise 5C

1   A particle collides normally with a fixed vertical wall. The two diagrams show the speeds of the particle before and after collision. In each case find the value of the coefficient of restitution e.

Before impact: After impact:

(a)

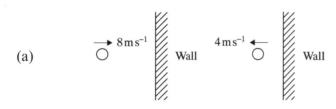

(b)

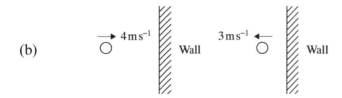

2   A particle collides normally with a fixed vertical wall. The two diagrams show the speed of the particle before impact. In case (a) the coefficient of restitution is $\frac{2}{3}$ and in case (b) the coefficient of restitution is $\frac{1}{4}$. Find the speed of the particle after impact in each case.

Before impact: After impact:

(a)

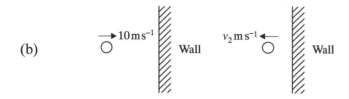

(b)

3 A particle collides normally with a fixed vertical wall. The two
diagrams show the speed of the particle after impact. In case (a)
the coefficient of restitution is $\frac{3}{5}$ and in case (b) the coefficient of
restitution is $\frac{1}{2}$. Find u_1 and u_2.

Before impact: After impact:

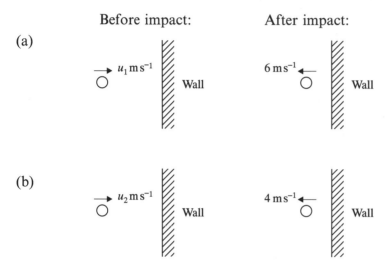

(a)

(b)

4 A small smooth ball of mass 0.4 kg is moving on a smooth
horizontal table with a speed of $12\,\mathrm{m\,s^{-1}}$ when it collides
normally with a fixed smooth vertical wall. It rebounds from
the wall with a speed of $9\,\mathrm{m\,s^{-1}}$.
  (a) Find the coefficient of restitution between the ball and the
      wall.
  (b) Find the kinetic energy lost by the ball due to the impact.
  (c) Describe a 'realistic situation' this could be used to model.
5 A small smooth ball strikes a smooth vertical wall at right
  angles. Its kinetic energy after impact is one-half of its initial
  kinetic energy. Find the coefficient of restitution between the
  ball and the wall.
6 A particle of mass m is travelling in a straight line with speed u
  on a smooth horizontal floor. It strikes a fixed smooth vertical
  wall normally. The kinetic energy lost by the particle due to the

collision is E. Show that the coefficient of restitution between the particle and the wall is given by:

$$\sqrt{\left(\frac{mu^2 - 2E}{mu^2}\right)}$$

7   A small ball falls from a height of 4 m on to a smooth horizontal plane. After rebounding from the plane it reaches a height of 1 m. Find the coefficient of restitution between the ball and the plane.

8   A particle falls from rest at a height of h metres on to a smooth horizontal plane. The coefficient of restitution between the particle and the plane is $\frac{1}{5}$. Given that the speed after the second impact is $2\,\mathrm{m\,s^{-1}}$ find the value of h to the nearest whole number.

9   A small smooth sphere falls from rest on to a smooth horizontal plane. It takes $1\frac{1}{4}$ seconds to reach the plane and another $\frac{3}{4}$ second to come to instantaneous rest. Find the coefficient of restitution between the sphere and the plane. What 'realistic situation' could this model?

10 A small smooth ball bearing falls from rest at a height H above a smooth horizontal plane. The coefficient of restitution between the ball bearing and the plane is $e\,(e < 1)$. Show that the total distance travelled by the ball bearing before coming to rest is:

$$\left(\frac{1 + e^2}{1 - e^2}\right)H$$

5.5 Successive impacts

The principles discussed in the previous sections may be applied to situations where three particles collide in pairs or where there are several collisions involving two particles and a wall. In such situations clear diagrams should be drawn showing the 'before' and 'after' information **for each collision**. This is illustrated in the next two examples.

Example 11

Three particles A, B and C have mass $1\,\text{kg}$, $2\,\text{kg}$ and $2\,\text{kg}$ respectively. They lie in a straight line with B between A and C. The coefficient of restitution between any pair of these spheres is e. Initially B and C are at rest and A is projected towards B with speed $6\,\text{m s}^{-1}$. After collision A rebounds from B with speed $0.5\,\text{m s}^{-1}$.

(a) Show that $e = \frac{5}{8}$.

(b) Find the speed of C after the second impact.

(a) *1st impact*

Before: $\longrightarrow 6\,\text{m s}^{-1}$ $\longrightarrow 0\,\text{m s}^{-1}$
$\bigcirc$ $\bigcirc$
$A\,(1\,\text{kg})$ $B\,(2\,\text{kg})$

After: $\bigcirc$ $\bigcirc$
$\longleftarrow$ $\longrightarrow$
$0.5\,\text{m s}^{-1}$ $u\,\text{m s}^{-1}$

Using the principle of conservation of momentum gives:
$$1 \times 6 + 2 \times 0 = 1 \times (-0.5) + 2u$$
so:
$$u = \tfrac{13}{4}$$

This is the speed of B after the 1st impact.

Newton's law of restitution gives:
$$e = \frac{u + 0.5}{6 - 0}$$

And substituting for u gives:
$$e = \frac{\frac{13}{4} + \frac{1}{2}}{6} = \frac{15}{4 \times 6}$$
$$= \frac{5}{8}$$

(b) *2nd impact*

Before: $\longrightarrow \frac{13}{4}\,\text{m s}^{-1}$ $\longrightarrow 0\,\text{m s}^{-1}$
$\bigcirc$ $\bigcirc$
$B\,(2\,\text{kg})$ $C\,(2\,\text{kg})$

After: $\bigcirc$ $\bigcirc$
$\longrightarrow v_1\,\text{m s}^{-1}$ $\longrightarrow v_2\,\text{m s}^{-1}$

Using the principle of conservation of momentum gives:
$$2 \times \frac{13}{4} + 2 \times 0 = 2v_1 + 2v_2$$

or:
$$v_1 + v_2 = \frac{13}{4} \tag{1}$$

Newton's law of restitution gives:
$$e = \frac{5}{8} = \frac{v_2 - v_1}{\frac{13}{4} - 0}$$

or:
$$v_2 - v_1 = \left(\frac{5}{8}\right)\left(\frac{13}{4}\right) = \frac{65}{32} \tag{2}$$

Eliminating v_1 and finding v_2 by adding equations (1) and (2):
$$2v_2 = \frac{65}{32} + \frac{13}{4}$$
$$= \frac{65 + 8 \times 13}{32} = \frac{169}{32}$$

So:
$$v_2 = \frac{169}{64} = 2\frac{41}{64}$$

The speed of C after the second impact is $2\frac{41}{64}\,\text{m s}^{-1}$.

Example 12

Particles A and B have mass 0.75 kg and 0.6 kg respectively. They are approaching each other moving along the same straight line with speeds of $6\,\text{m s}^{-1}$ and $4\,\text{m s}^{-1}$ respectively. The coefficient of restitution between them is $\frac{4}{5}$. After the collision the direction of motion of B is reversed. Particle B then strikes a fixed vertical wall at right angles. If the coefficient of restitution between B and the wall is $\frac{2}{3}$, show that B will collide again with A and find the speeds of A and B after this second collision.

1st impact

Before:

    $\longrightarrow 6\,\text{m s}^{-1}$ $4\,\text{m s}^{-1} \longleftarrow$

    ◯   ◯

  A (0.75 kg) B (0.6 kg) $\left(e = \frac{4}{5}\right)$

After:

    ◯   ◯

    $\longrightarrow v_1\,\text{m s}^{-1}$ $\longrightarrow v_2\,\text{m s}^{-1}$

Using conservation of momentum gives:
$$(0.75) \times 6 - (0.6) \times 4 = (0.75)v_1 + (0.6)v_2$$

or:
$$5v_1 + 4v_2 = 14 \tag{1}$$

Newton's Law of Restitution gives:
$$e = \frac{4}{5} = \frac{v_2 - v_1}{6 + 4}$$

or: $$v_2 - v_1 = 10 \times \frac{4}{5} = 8 \qquad (2)$$

Solving equations (1) and (2) simultaneously gives:

$$v_1 = -2 \quad \text{and} \quad v_2 = 6$$

So after the 1st impact:

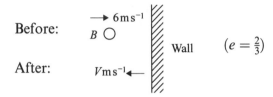

taking account of the minus sign in v_1.

2nd impact

Before: → 6 m s⁻¹
           B ○ | Wall $(e = \frac{2}{3})$

After: V m s⁻¹ ←

By Newton's law of restitution:

$$e = \frac{V}{v} \quad \text{or} \quad V = ev$$

So: $$V = \frac{2}{3} \times 6 = 4$$

The rebound speed of B is $4\,\mathrm{m\,s^{-1}}$.

2nd collision between A and B

2 m s⁻¹ ←○ A   4 m s⁻¹ ←○ B $(e = \frac{4}{5})$

As B is moving faster than A there will be a further collision between A and B.

Before: 2 m s⁻¹ ←   4 m s⁻¹ ←
           ○   ○
           A (0.75 kg) B (0.6 kg)

After: → V_1 m s⁻¹   → V_2 m s⁻¹

Using conservation of momentum, taking the direction of V_1 and V_2 in the diagram as positive, gives:

$$-(0.75) \times 2 - (0.6)4 = (0.75)V_1 + (0.6)V_2$$

or: $$-26 = 5V_1 + 4V_2 \qquad (3)$$

Newton's law of restitution gives:

$$e = \frac{4}{5} = \frac{V_2 - V_1}{4 - 2}$$

or:

$$V_2 - V_1 = 2 \times \frac{4}{5} = \frac{8}{5} \tag{4}$$

Eliminating V_1 between equations (3) and (4) gives:

$$-18 = 9V_2$$

So:

$$V_2 = -2$$

Substituting this value of V_2 into equation (4) gives:

$$V_1 = -3.6$$

So after this collision:

The speeds of A and B after the second collision are $3.6\,\mathrm{m\,s^{-1}}$ and $2\,\mathrm{m\,s^{-1}}$ respectively.

Exercise 5D

1 Each part of this question involves 3 small smooth spheres A, B and C of equal radius moving along the same straight line on a smooth horizontal plane. The sphere A collides with B, and then B collides with C. The diagrams show the situations

(i) before any collision,

(ii) after A and B have collided,

(iii) after B has collided with C.

In (a) $e = \frac{1}{2}$ between any two spheres in collision, and in

(b) $e = \frac{3}{4}$ between any two spheres in collision.

In each part find x, y, v and w.

2 A small smooth sphere *A* of mass 2 kg moving on a smooth horizontal floor with speed 5 m s^{-1} collides directly with a small smooth sphere of mass 1 kg which is at rest. The coefficient of restitution between the spheres is $\frac{4}{5}$.

(a) Find the speed of *B* after the impact.

Following the impact sphere *B* goes on to strike a vertical wall which is at right angles to the direction of motion of the spheres. The coefficient of restitution between sphere *B* and the wall is *e*. The sphere *A* is brought to rest by a second collision with *B*.

(b) Show that $e = \frac{2}{9}$.

3 Three small smooth spheres *A*, *B* and *C* have mass 0.1 kg, 0.2 kg and 0.4 kg respectively. They are at rest on a smooth table with their centres collinear and *B* between *A* and *C*. Sphere *A* is projected directly towards *B* with speed 8 m s^{-1}. After *C* has been struck by *B* it moves with a speed of 2 m s^{-1}. The coefficient of restitution between any two spheres is *e*.

(a) Show that $e = \frac{1}{2}$.

(b) Show that *A* and *B* are brought to rest by the collision.

4 Three perfectly elastic particles *A*, *B* and *C* with masses 3 kg, 2 kg and 1 kg respectively lie at rest in a straight line on a smooth horizontal table with *B* between *A* and *C*. Particle *A* is projected directly towards *B* with speed 5 m s^{-1} and after *A* has collided with *B*, *B* collides with *C*.

(a) Find the speed of each particle after the second impact.

(b) State, giving a reason, whether there will be a third collision.

5 Three small smooth identical spheres each of mass 1 kg move on a straight line on a smooth horizontal floor. Initially *B* lies between *A* and *C*. Spheres *A* and *B* are projected directly towards each other with speeds 3 m s^{-1} and 2 m s^{-1} respectively and sphere *C* is projected directly away from sphere *A* with speed 2 m s^{-1}. The coefficient of restitution between any two spheres is *e*. Show that sphere *B* will only collide with sphere *C* if $e > \frac{3}{5}$. Describe a 'realistic situation' this could be used to model.

6 Three small smooth spheres *A*, *B* and *C*, of equal size and of mass 2*m*, 7*m* and 14*m* respectively, are at rest on a smooth

horizontal floor with their centres in that order on a straight line. The coefficient of restitution between each pair of spheres is $\frac{1}{2}$. Sphere A is projected so that it collides with sphere B.

(a) Show that, after two collisions have taken place, sphere B is at rest and spheres A and C are moving in opposite directions with equal speeds.

(b) Find what fraction of the original kinetic energy remains after the two collisions.

7 Two small smooth balls A and B of equal radius and mass $4\,\text{kg}$ and $5\,\text{kg}$ respectively lie at rest on a smooth horizontal floor. Ball A is projected with speed u and collides with ball B. Following this collision ball B then strikes a smooth vertical wall normally. After rebounding from the wall, ball B again collides with ball A. Given that ball B is brought to rest by this second collision with A,

(a) show that $2e^3 - 3e^2 - 3e + 2 = 0$,

where e is the coefficient of restitution between the two balls and between ball B and the wall.

(b) verify that $e = \frac{1}{2}$ is the only practical solution of the equation.

SUMMARY OF KEY POINTS

1 **Conservation of linear momentum**

When two particles collide:

$$\frac{\text{total momentum}}{\text{before the collision}} = \frac{\text{total momentum}}{\text{after the collision}}$$

$$m_1 u_1 + m_2 u_2 = m_1 v_1 + m_2 v_2$$

Before collision: m_1 ○ m_2 ○
    → u_1 → u_2

After collision: → v_1 → v_2

2 **Newton's law of restitution**

$$\frac{\text{speed of separation of particles}}{\text{speed of approach of particles}} = e$$

where e is the coefficient of restitution between the particles.

3 Impact of a particle normally with a fixed surface.

Before impact:

After impact:

(speed of rebound) $= e$ (speed of approach)

where e is the coefficient of restitution between the particle and the surface.

Statics of rigid bodies

6

6.1 Centre of mass of a uniform plane lamina

In chapter 6 of Book M1 the centre of mass of a uniform plane lamina was determined using its symmetries. Now a more general method of determining the position of the centre of mass is required which can be used when the lamina has no symmetries. This more general method uses the ideas of calculus.

The following 'key points' from chapter 6 of Book M1, will be required:

(1) The centre of mass of a lamina is the point at which the weight acts.

(2) The weight of a uniform lamina is evenly distributed throughout its area.

(3) The centre of mass of a lamina must lie on any axis of symmetry.

(4) The centre of mass of a system of n masses m_1, m_2, ..., m_n at points with coordinates (x_1, y_1), (x_2, y_2), ..., (x_n, y_n) has coordinates $(\bar{x}, \bar{y})$

where: $\qquad \bar{x} = \dfrac{\sum m_i x_i}{\sum m_i} \qquad$ and $\qquad \bar{y} = \dfrac{\sum m_i y_i}{\sum m_i}$

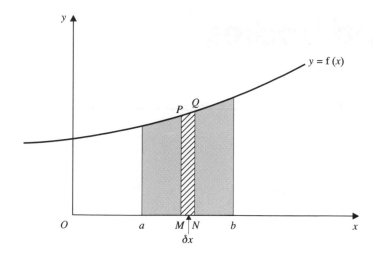

Consider a uniform lamina bounded by the curve with equation
$y = f(x)$, the x-axis and the lines $x = a$ and $x = b$ (the region lightly
shaded in the figure). Let the mass per unit area of the lamina be ρ
(the Greek lower case letter 'rho'). Consider the lamina as made up
of elementary strips such as $MNQP$ each parallel to the y-axis. Let
$OM = x$ and $MN = \delta x$.

Then:

length of elementary strip $MNQP = y$

area of elementary strip $= y\delta x$

mass of elementary strip $= \rho y \delta x$

coordinates of centre of mass of strip $\simeq (x, \tfrac{1}{2} y)$

The x coordinate of the centre of mass of the lamina, $\bar{x}$, is now
obtained from:

$$\bar{x} = \frac{\sum_i m_i x_i}{\sum_i m_i} \simeq \frac{\sum_{x=a}^{x=b} (\rho y \delta x) x}{\sum_{x=a}^{x=b} (\rho y \delta x)}$$

Taking the limit as $\delta x \to 0$ the summations become integrals. So for
the lamina shown in the figure:

■
$$\bar{x} = \frac{\int_a^b x(\rho y \, dx)}{\int_a^b \rho y \, dx} = \frac{\int_a^b x f(x) \, dx}{\int_a^b f(x) \, dx}$$

To find the y coordinate of the centre of mass of the lamina it is
necessary to note that the centre of mass of the strip has y
coordinate equal to $\tfrac{1}{2} y$. Then:

- $$\bar{y} = \frac{\int_a^b \frac{1}{2}y(\rho y \, \mathrm{d}x)}{\int_a^b \rho y \, \mathrm{d}x} = \frac{\frac{1}{2}\int_a^b y^2 \, \mathrm{d}x}{\int_a^b y \, \mathrm{d}x} \qquad \text{where } y = f(x).$$

These formulae are often written in the alternative form:

- $$\left[\int_a^b \rho y \, \mathrm{d}x\right] \bar{x} = M\bar{x}$$

$$= \int_a^b \rho xy \, \mathrm{d}x = \int_a^b \rho x \, f(x) \, \mathrm{d}x$$

$$\left[\int_a^b \rho y \, \mathrm{d}x\right] \bar{y} = M\bar{y}$$

$$= \frac{1}{2}\int_a^b \rho y^2 \, \mathrm{d}x = \frac{1}{2}\int_a^b [f(x)]^2 \, \mathrm{d}x$$

where M is the total mass of the lamina.

Example 1

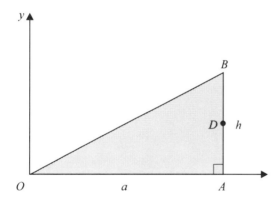

Find the coordinates of the centre of mass of the triangle OAB shown in the figure where $OA = a$ and $AB = h$.

The equation of OB is: $\qquad\qquad\qquad y = \dfrac{h}{a}x$

The mass of the triangle is: $\qquad \rho \, (\text{area}) = \rho\left(\frac{1}{2}ah\right) = M$

Using the formula for $\bar{x}$:

$$M\bar{x} = \int_0^a \rho xy \, \mathrm{d}x = \int_0^a \rho x \frac{h}{a}x \, \mathrm{d}x = \frac{h}{a}\rho \int_0^a x^2 \, \mathrm{d}x$$

$$= \frac{h}{a}\rho\left[\frac{x^3}{3}\right]_0^a = \frac{1}{3}ha^2\rho$$

So:
$$\bar{x} = \frac{\frac{1}{3}ha^2\rho}{\frac{1}{2}ha\rho} = \frac{2}{3}a$$

Similarly :
$$M\bar{y} = \int_0^a \rho \frac{1}{2}y^2 dx = \frac{1}{2}\rho \int \frac{h^2 x^2}{a^2}\, dx$$
$$= \frac{\rho h^2}{2a^2}\left[\frac{x^3}{3}\right]_0^a = \frac{\rho h^2 a}{6}$$

So:
$$\bar{y} = \frac{\frac{1}{6}\rho h^2 a}{\frac{1}{2}\rho ha} = \frac{1}{3}h$$

The centre of mass G of the triangle has coordinates $(\frac{2}{3}a, \frac{1}{3}h)$.

This shows that if D is the mid-point of AB then $GD = \frac{1}{3}OD$ as stated in section 6.4 of Book M1.

Example 2
Find the coordinates of the centre of mass of the uniform lamina bounded by the curve with equation $y^2 = 4x$, the x-axis and the line $x = 1$.

First sketch the lamina:

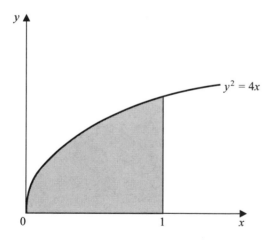

The mass of the lamina is:
$$M = \int_0^1 \rho y\, dx = \rho \int_0^1 2x^{\frac{1}{2}} dx$$
$$= 2\rho\left[\frac{2}{3}x^{\frac{3}{2}}\right]_0^1 = 2\rho \times \frac{2}{3} = \frac{4}{3}\rho$$

where ρ is the mass per unit area of the lamina.

The x coordinate of the centre of mass $\bar{x}$ is obtained from:

$$M\bar{x} = \int_0^1 \rho x y \, dx = \rho \int_0^1 x 2x^{\frac{1}{2}} \, dx = 2\rho \int_0^1 x^{\frac{3}{2}} \, dx$$

$$= 2\rho \left[\tfrac{2}{5} x^{\frac{5}{2}} \right]_0^1 = \tfrac{4}{5} \rho$$

So:
$$\bar{x} = \frac{\frac{4}{5}\rho}{\frac{4}{3}\rho} = \tfrac{3}{5}$$

To find $\bar{y}$ use:

$$M\bar{y} = \int_0^1 \rho \tfrac{1}{2} y^2 \, dx = \tfrac{1}{2}\rho \int_0^1 4x \, dx$$

$$= 2\rho \left[\frac{x^2}{2} \right]_0^1 = 2\rho \tfrac{1}{2} = \rho$$

So:
$$\bar{y} = \frac{\rho}{\frac{4}{3}\rho} = \tfrac{3}{4}$$

The coordinates of the centre of mass are $(\tfrac{3}{5}, \tfrac{3}{4})$.

Example 3

Show that the centre of mass of a uniform semicircular lamina lies on its axis of symmetry at a distance of $\dfrac{4r}{3\pi}$ from the bounding diameter.

Take the bounding diameter as the y-axis and the centre O as the origin. If the point A, on the curved boundary, has coordinates (x, y) then by Pythagoras' theorem $x^2 + y^2 = r^2$.

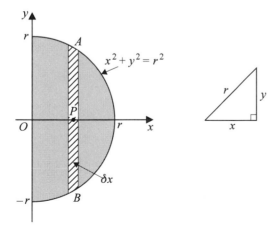

Consider the elementary strip AB of width δx, parallel to the y-axis and at a distance of x from it. The centre of mass of this strip is at the point P with coordinates $(x, 0)$.

The centres of mass of all such strips also lie on the x-axis.

Hence the centre of mass of the lamina will also lie on the x-axis – the axis of symmetry of the lamina.

By Pythagoras' theorem:

$$AP = \sqrt{(r^2 - x^2)}$$

The area of the strip AB is therefore $2\sqrt{(r^2 - x^2)}\delta x$ and its mass is $2\rho\sqrt{(r^2 - x^2)}\delta x$ where ρ is the mass per unit area of the lamina.

The mass of the lamina is $M = \frac{1}{2}\pi r^2\rho$ and $\bar{x}$ is obtained from:

$$\begin{aligned}
M\bar{x} &= \int_0^r x2\rho\sqrt{(r^2 - x^2)}\,\mathrm{d}x \\
&= 2\rho\int_0^r x(r^2 - x^2)^{\frac{1}{2}}\,\mathrm{d}x \\
&= 2\rho\left[-\tfrac{2}{6}\left(r^2 - x^2\right)^{\frac{3}{2}}\right]_0^r \\
&= \tfrac{2}{3}\rho r^3
\end{aligned}$$

So:

$$\bar{x} = \frac{\frac{2}{3}\rho r^3}{\frac{1}{2}\rho\pi r^2} = \frac{4r}{3\pi}$$

That is, the centre of mass is at a distance $\dfrac{4r}{3\pi}$ from the bounding diameter.

Example 4

Show that the centre of mass of a uniform wire in the form of an arc of a circle, radius r, subtending an angle 2α at the centre is at a distance $\dfrac{r \sin \alpha}{\alpha}$ from the centre.

The bisector of the angle 2α is an axis of symmetry of the wire. **The centre of mass will therefore lie on this line**. Take this line to be the x-axis and the centre O of the circle to be the origin.

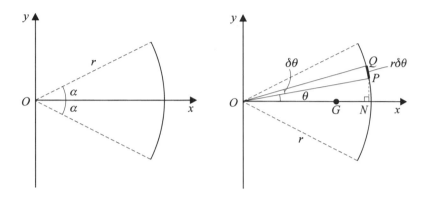

Let ω be the mass per unit length of the wire.

A typical element PQ of the wire subtends an angle $\delta\theta$ at O and has length $r\delta\theta$.

The mass of this element is $\omega r\delta\theta$ and from the diagram the length ON is $r\cos\theta$. (Here N is the foot of the perpendicular from P on to Ox).

The length of the wire is $r2\alpha$ (see chapter 7 of Book P1) so the mass M of the wire is

$$M = (r2\alpha)\omega$$

The x coordinate, $\bar{x}$, of the centre of mass is obtained from:

$$M\bar{x} = \sum_{\theta=-\alpha}^{\theta=\alpha} (\omega r\delta\theta) r \cos\theta$$

Taking the limit as $\delta\theta \to 0$ the summation becomes an integral and so:

$$M\bar{x} = \omega r^2 \int_{-\alpha}^{\alpha} \cos\theta \, \delta\theta = \omega r^2 [\sin\theta]_{-\alpha}^{\alpha}$$
$$= 2\omega r^2 \sin\alpha$$

and:
$$\bar{x} = \frac{2wr^2 \sin \alpha}{2wr\alpha} = \frac{r \sin \alpha}{\alpha}$$

The centre of mass of the wire is at a distance $\dfrac{r \sin \alpha}{\alpha}$ from the centre and lies on the axis of symmetry.

Example 5

Find the centre of mass of a uniform lamina in the form of a sector of a circle, radius r, subtending an angle 2α at the centre.

As in the previous example the bisector of the angle 2α is an axis of symmetry. Choose axes as before with the centre O as the origin and the line of symmetry as the x-axis.

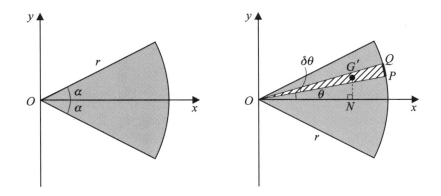

Let ρ be the mass per unit area.

Divide the lamina into elements of the form OPQ, which is a sector subtending an angle of $\delta\theta$ at O (see diagram).

The area of OPQ is $\frac{1}{2}r^2\delta\theta$. (See chapter 7 of Book P1.) Therefore the mass of OPQ is $\frac{1}{2}\rho r^2\delta\theta$.

The sector OPQ is approximately a triangle, so its centre of mass G' is at a distance $\frac{2}{3}r$ from O so that $ON = \frac{2}{3}r\cos\theta$ (see diagram).

The mass M of the whole sector is given by:

$$M = \rho \times \text{(area of whole sector)}$$
$$= \tfrac{1}{2}\rho r^2 2\alpha = \rho r^2 \alpha$$

The x coordinate, $\bar{x}$, of the centre of mass is obtained from:

$$M\bar{x} = \sum_{\theta=-\alpha}^{\theta=\alpha} \left(\tfrac{1}{2}\rho r^2 \delta\theta\right)\tfrac{2}{3}r\cos\theta$$

Taking the limit as $\delta\theta \to 0$ the summation becomes an integral and so:

$$M\bar{x} \int_{-\alpha}^{+\alpha} \tfrac{1}{3}\rho r^3 \cos\theta \, d\theta = \tfrac{1}{3}\rho r^3 [\sin\theta]_{-\alpha}^{\alpha}$$
$$= \tfrac{1}{3}\rho r^3 2\sin\alpha = \tfrac{2}{3}\rho r^3 \sin\alpha$$

and:

$$\bar{x} = \frac{\tfrac{2}{3}\rho r^3 \sin\alpha}{\rho r^2 \alpha} = \frac{2r\sin\alpha}{3\alpha}$$

The centre of mass is at a distance $\dfrac{2r\sin\alpha}{3\alpha}$ from the centre of the sector and lies on the axis of symmetry.

In the following example a lamina with a more complicated shape is considered. Such a lamina may be considered as a combination of several simple laminae such that:

(i) the mass of each is easily obtained,

(ii) the position of the centre of mass of each is given in the list of standard results at the end of this chapter and also in the formula book.

If integration is not explicitly asked for in the question you may, in general, quote these standard results.

Example 6

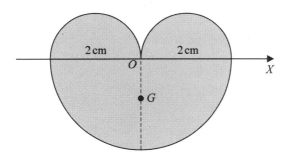

The figure shows a badge cut from a uniform sheet of metal. The badge is formed from two semicircles, each of radius 1 cm and a semicircle of radius 2 cm joined as shown in the figure to make a plane lamina. The point O is the point of contact of the smaller semicircles. Determine, in terms of π, the distance from O of the centre of mass.

The line through O perpendicular to the joining line OX is a line of symmetry and so the centre of mass G lies on this line. Let the mass per unit area be ρ. Then:

mass of smaller semicircle is $\quad \frac{1}{2}\pi(1)^2\rho = \frac{1}{2}\pi\rho$

mass of larger semicircle is $\quad \frac{1}{2}\pi(2)^2\rho = 2\pi\rho$

So: mass of badge is $\quad 2\pi\rho + 2(\frac{1}{2}\pi\rho) = 3\pi\rho$

Using the standard result for the centre of mass of a semicircular lamina, $\frac{4r}{3\pi}$ (see the table at the end of this chapter or the formula book), the centre of mass of a smaller semicircle is a distance $\frac{4}{3\pi}$ **above** OX.

The centre of mass of the larger semicircle is a distance $\frac{4 \times 2}{3\pi} = \frac{8}{3\pi}$ **below** OX.

Let the centre of mass of the badge be a distance $\bar{x}$ **below** OX. Then, as in chapter 6 of Book M1,

$$3\pi\rho\bar{x} = 2\pi\rho\left(\frac{8}{3\pi}\right) - \frac{1}{2}\pi\rho\left(\frac{4}{3\pi}\right) - \frac{1}{2}\pi\rho\left(\frac{4}{3\pi}\right)$$

So: $$\bar{x} = \frac{4}{3\pi}$$

The distance from O of the centre of mass is $\frac{4}{3\pi}$.

Exercise 6A

1 Find the centre of mass of a uniform triangular lamina bounded by the lines $3y = 4x$, $y = 0$ and $x = 3$.

2 Use integration to find the centre of mass of a uniform lamina bounded by the curve with equation $y = x^2$, the x-axis and the line $x = 1$.

3 Use integration to find the centre of mass of a uniform lamina bounded by the curve with equation $y = x^3$, the x-axis and the lines $x = 1$ and $x = 2$.

4 Use integration to find the centre of mass of a uniform lamina bounded by the curve with equation $y^2 = 4ax$ and the line $x = b \, (b > 0)$.

5 Use integration to find the centre of mass of a uniform triangular lamina with vertices $(0, h)$, $(0, 0)$ and $(a, 0)$.

6 Find the centre of mass of a uniform lamina bounded by the curve with equation $y = x^2 + 1$, the x-axis and the lines $x = 0$ and $x = 1$.

7

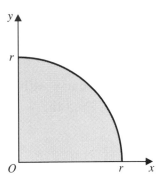

Find the centre of mass of a uniform lamina in the shape of the quadrant of a circle of radius r shown in the figure.

8 Find the centre of mass of a uniform lamina bounded by the curve with equation $y = \cos x \, (x > 0)$ and the lines $x = 0$ and $y = 0$.

9 Find the centre of mass of a uniform lamina bounded by the curve with equation $y = e^x$, the x-axis and the lines $x = 0$ and $x = 1$.

In questions 10 and 11 use the 'subtraction method', that is consider the given lamina as the difference between two laminae.

10

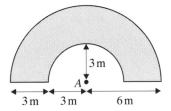

The figure shows a uniform semicircular lamina of radius 6 m from which a semicircular portion of radius 3 m has been removed. Find the distance of the centre of mass of the remaining shape from the point A shown in the figure.

11

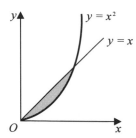

The figure shows a uniform lamina bounded by the curve with equation $y = x^2$ and the line with equation $y = x$. Find the coordinates of the centre of mass of the lamina.

6.2 Centre of mass of a uniform solid body

For a uniform solid body the 'key points' from chapter 6 of Book M1 for a uniform lamina may be generalised so that:

(1) The centre of mass of a solid body is the point at which the weight acts.

(2) The weight of a uniform solid body is evenly distributed throughout its volume.

(3) (a) The centre of mass of a uniform solid body must lie on any **axis** of symmetry.
   (b) The centre of mass of a uniform solid body must lie on any **plane** of symmetry.

(4) The centre of mass of a system of n masses m_1, m_2, ..., m_n at points with coordinates (x_1, y_1, z_1), (x_2, y_2, z_2), ..., (x_n, y_n, z_n) has coordinates $(\bar{x}, \bar{y}, \bar{z})$ where:

$$\bar{x} = \frac{\sum m_i x_i}{\sum m_i}$$

$$\bar{y} = \frac{\sum m_i y_i}{\sum m_i}$$

$$\bar{z} = \frac{\sum m_i z_i}{\sum m_i}$$

Use of symmetry

The position of the centre of mass of certain uniform solids may be obtained using just the symmetry properties.

Uniform solid sphere

A uniform solid sphere has an infinite number of planes of symmetry – every plane through the centre of the sphere is a plane of symmetry. Since each of these contains the centre G this is therefore the centre of mass.

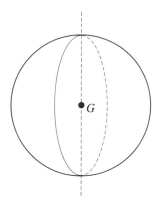

Uniform solid right circular cylinder

The cross-section of a solid right circular cylinder is a circle. The cylinder therefore has an infinite number of planes of symmetry, each of which contains the axis of the cylinder. Hence the centre of mass of a uniform right circular cylinder must lie on the axis.

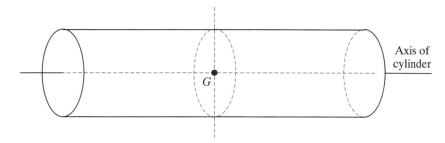

The cylinder also has a plane of symmetry which bisects the axis of the cylinder and is parallel to the plane ends of the cylinder. The centre of mass must also lie on this plane. This plane intersects the axis of symmetry at the point G mid-way between the ends of the cylinder. The point G is the centre of mass of the cylinder.

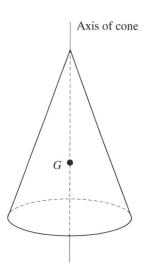

Uniform solid right circular cone

The axis of a solid right circular cone is an axis of symmetry of the cone and every cross-section perpendicular to this axis is a circle with centre on this line. Hence the centre of mass G of a uniform circular cone lies on the axis of the cone. The exact position of G on the axis will be determined later in this chapter.

Uniform solid hemisphere
Suppose O is the centre of the plane face of the hemisphere. The line through O perpendicular to the plane face is an axis of symmetry. Hence the centre of mass of a uniform solid hemisphere lies on this line. The exact position of G on this line will be found later in this chapter.

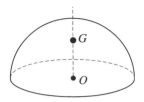

Centres of mass of solids of revolution

Finding the centre of mass of an irregular solid is very difficult. An experimental method may be the only possible approach. There is however a special type of solid known as a **solid of revolution** for which there is a general method using calculus.

A solid of revolution is formed by rotating a region bounded by a known curve through 360° about a fixed axis. Examples of such solids are:

hemisphere, obtained by rotating a quarter circle

right circular cone, obtained by rotating a triangle

paraboloid, obtained by rotating a parabola.

As in section 6.1, consider the region bounded by the curve with equation $y = f(x)$, the x-axis and the lines $x = a$ and $x = b$. This is rotated through 360° about the x-axis to produce a solid of revolution bounded by $x = a$ and $x = b$. A cross-section of the solid produced is shown in the figure.

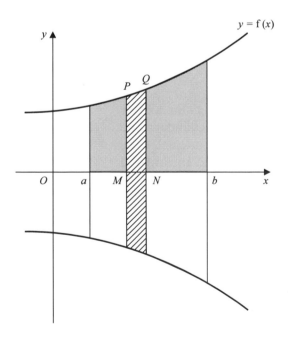

Consider a typical element of the solid formed by rotating the strip $MNQP$ about the x-axis where $OM = x$ and $MN = \delta x$. This typical element is a very thin cylinder whose cross-section is a circle of radius $MP = y$ and area πy^2

So: $$\text{volume of element} = \pi y^2 \delta x$$

$$\text{mass of element} = \rho \pi y^2 \delta x$$

where ρ is the mass per unit volume.

By symmetry the centre of mass of this element lies on the x-axis. Its x coordinate is equal to $OM = x$.

The x coordinate of the centre of mass of the whole solid, $\bar{x}$, is now obtained from:

$$\bar{x} = \frac{\sum m_i x_i}{\sum m_i} = \frac{\sum\limits_{x=a}^{x=b} (\rho \pi y^2 \delta x) x}{\sum\limits_{x=a}^{x=b} (\rho \pi y^2 \delta x)}$$

Taking the limit as $\delta x \to 0$, the summations become integrals. So for the solid shown in the figure:

$$\bar{x} = \frac{\int_a^b x \rho \pi y^2 \, dx}{\int_a^b \rho \pi y^2 \, dx}$$

or:

$$M\bar{x} = \left[\int_a^b \rho \pi y^2 \, dx\right] \bar{x} = \int_a^b \rho \pi y^2 x \, dx$$

Example 7

A uniform solid right circular cone has base radius a and height h. Show that the centre of mass lies on the axis of the cone at a distance $\dfrac{h}{4}$ from the plane base.

Assume the axis of symmetry of the cone is the x-axis. The solid right circular cone may be generated by rotating a triangular region about the x-axis as shown in the figure.

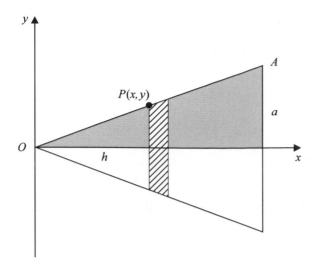

To find the equation of the line OA let the point $P(x, y)$ be on this line. The gradient of the line is $\dfrac{a}{h}$ and so:

$$\frac{y}{x} = \frac{a}{h}$$

or:

$$y = \frac{ax}{h}$$

The shaded triangular region is therefore bounded by the line $y = \dfrac{ax}{h}$, the x-axis and the line $x = h$.

The mass M of the cone is given by:

$$M = \text{volume of cone} \times \rho = \tfrac{1}{3}\pi a^2 h \rho$$

Then:

$$M\bar{x} = \int_0^h x\rho\pi y^2 \,\mathrm{d}x \quad \text{with } y = \frac{a}{h}x$$

$$= \pi\rho \int_0^h x \frac{a^2 x^2}{h^2} \,\mathrm{d}x = \pi \frac{\rho a^2}{h^2} \int_0^h x^3 \,\mathrm{d}x$$

$$= \pi \frac{\rho a^2}{h^2} \left[\frac{x^4}{4}\right]_0^h = \pi \frac{\rho a^2}{h^2}\frac{h^4}{4} = \tfrac{1}{4}\pi\rho a^2 h^2$$

So:

$$\bar{x} = \frac{\tfrac{1}{4}\pi\rho a^2 h^2}{\tfrac{1}{3}\pi\rho a^2 h} = \tfrac{3}{4}h$$

and the centre of mass is a distance of $\frac{3}{4}h$ from O or $\frac{1}{4}h$ from the plane base. As discussed earlier, the centre of mass lies on the axis of the cone as this is an axis of symmetry.

Example 8

Show that the centre of mass of a solid hemisphere of radius a is at a distance $\dfrac{3a}{8}$ from its plane face.

The hemisphere may be obtained by rotating a quarter circle of radius a and centre the origin O, in the 1st quadrant, about the x-axis.

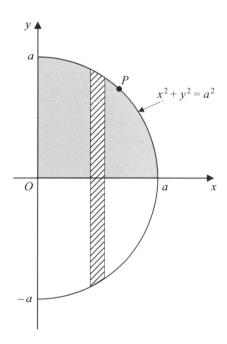

If the coordinates of any point P on the curved boundary are (x, y) then by Pythagoras' theorem $x^2 + y^2 = a^2$. The shaded region is bounded by this curve, the x-axis and the y-axis.

The mass M of the hemisphere is given by:

$$M = \text{volume of hemisphere} \times \rho = \left(\tfrac{2}{3}\pi a^3\right)\rho$$

Then:
$$M\bar{x} = \int_0^a x\rho\pi y^2 \, dx \qquad \text{with } y^2 = a^2 - x^2$$

$$= \pi\rho \int_0^a x(a^2 - x^2) \, dx = \pi\rho \int_0^a (a^2 x - x^3) \, dx$$

$$= \pi\rho \left[a^2 \frac{x^2}{2} - \frac{x^4}{4} \right]_0^a$$

$$= \pi\rho \left[\frac{a^4}{2} - \frac{a^4}{4} \right] = \frac{\pi}{4}\rho a^4$$

So:
$$\bar{x} = \frac{\frac{\pi}{4}\rho a^4}{\frac{2\pi}{3}\rho a^3} = \frac{3a}{8}$$

The centre of mass is at a distance $\dfrac{3a}{8}$ from the plane face.

Example 9

The region R is bounded by the curve with equation $y^2 = x$, the x-axis and the lines $x = 2$, $x = 4$, and lies in the 1st quadrant. The region is rotated about the x-axis through one revolution. Find the distance of the centre of mass of the solid of revolution from O.

The region R is shown shaded in the figure together with the solid of revolution formed.

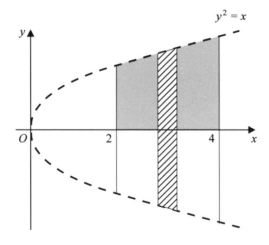

The x-axis is an axis of symmetry of the solid and so the centre of mass lies on this axis.

The mass of the solid of revolution M is given by:

$M = \rho \times$ (volume of solid of revolution)

$$M = \rho \int_2^4 \pi y^2 \, dx \qquad \text{where } y^2 = x \text{ (See chapter 7 of Book P2)}$$

$$= \rho\pi \int_2^4 x \, dx = \rho\pi \left[\frac{x^2}{2}\right]_2^4$$

$$= \rho\pi \left[\frac{(4)^2}{2} - \frac{(2)^2}{2}\right] = \rho\pi[8 - 2]$$

$$= 6\rho\pi$$

Then: $\qquad M\bar{x} = \int_2^4 x\rho\pi y^2 \, dx \qquad \text{where } y^2 = x$

$$= \pi\rho \int_2^4 xx \, dx = \pi\rho \int_2^4 x^2 \, dx$$

$$= \pi\rho \left[\frac{x^3}{3}\right]_2^4 = \frac{\pi\rho}{3}\left[(4)^3 - (2)^3\right] = \frac{\pi\rho}{3}[64 - 8]$$

$$= \frac{56}{3}\pi\rho$$

So: $\qquad\qquad\qquad \bar{x} = \dfrac{56\pi\rho}{3(6\pi\rho)} = \dfrac{56}{18} = \dfrac{28}{9} = 3\dfrac{1}{9}$

The centre of mass is at a distance $3\frac{1}{9}$ from O on the axis of symmetry.

Centres of mass of surfaces of revolution

When part of a **curve** is rotated through $360°$ about a fixed line we obtain a surface known as a **surface of revolution**. Examples of such surfaces are a hollow cone and a hemispherical shell.

The centre of mass of a surface of revolution is in general difficult to obtain but we can obtain its position for a hemispherical shell by using calculus methods similar to those used for a volume of revolution.

The hemispherical shell

The hemispherical shell may be obtained by rotating the quarter of the circle $x^2 + y^2 = a^2$ in the 1st quadrant about the x-axis as shown in the figure.

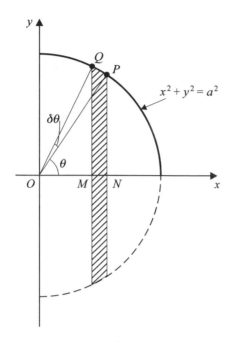

Consider the shaded strip obtained by rotating the arc PQ of the circle about the x-axis. Since the circle is of radius a the length of arc PQ is $a\delta\theta$, where $\angle POQ = \delta\theta$ and $\angle xOP = \theta$.

The shaded strip is approximately the curved surface of a cylinder of radius NP and height $a\delta\theta$. So the area is:

$$2\pi(NP)(a\delta\theta) = 2\pi a \sin\theta \, a\delta\theta$$

Therefore the mass of the shaded strip is:

$$2\pi\rho a^2 \sin\theta \, \delta\theta$$

where ρ is the mass per unit area.

Hence the total mass of the hemispherical shell, M, is:

$$M = 2\pi \int_0^{\frac{\pi}{2}} \rho a^2 \sin\theta \, d\theta = 2\pi\rho a^2 [-\cos\theta]_0^{\frac{\pi}{2}}$$
$$= 2\pi\rho a^2$$

[You could obtain the same result using the formula $4\pi a^2$ for a sphere, dividing by 2 to obtain the surface area of a hemisphere and then multiplying by ρ.]

Using $M\bar{x} = \sum_i m_i x_i$ with $m_i = 2\pi\rho a^2 \sin\theta \, d\theta$ and $x_i = ON = a\cos\theta$

then:
$$M\bar{x} = \int_0^{\frac{\pi}{2}} (2\pi\rho a^2 \sin\theta)\, d\theta\, a\cos\theta$$

So:
$$M\bar{x} = 2\pi\rho a^3 \int_0^{\frac{\pi}{2}} \sin\theta \cos\theta\, d\theta$$

$$= \pi\rho a^3 \int_0^{\frac{\pi}{2}} \sin 2\theta\, d\theta$$

since $\sin\theta\cos\theta = \frac{1}{2}\sin 2\theta$

Integrating:
$$M\bar{x} = \pi\rho a^3 \left[-\tfrac{1}{2}\cos 2\theta \right]_0^{\frac{\pi}{2}}$$

$$= \pi\rho a^3$$

So:
$$\bar{x} = \frac{\pi\rho a^3}{2\pi\rho a^2} = \frac{a}{2}$$

Hollow cone

To obtain the centre of mass for a hollow cone a rather different method is used. Consider the surface of the cone to be made up of elementary strips each of which is a triangle with one vertex the vertex V of the cone as in the figure.

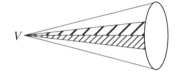

As shown in example 1 of this chapter, the centre of mass of each triangle, which can be considered as isosceles, is on the axis of symmetry and $\frac{2}{3}$ of the distance from the vertex V to the opposite side. So the centre of mass of the hollow cone is also two-thirds of the way from the vertex V to the base and by symmetry will be on the axis of the cone. It is more usual to say that for a hollow cone of height h the centre of mass is on the axis and $\frac{1}{3}h$ from the base.

Composite bodies

In chapter 6 of Book M1 a composite plane figure was considered as a combination of simple plane laminae, the positions of whose centres of mass were known. Composite bodies may be considered in the same way and this is illustrated in the following example.

Example 10

A uniform solid cone of height h and base radius r is joined to a uniform solid hemisphere of radius r so that the circular faces of the bodies coincide. The cone and the hemisphere have the same density. Find the distance of the centre of mass G of this composite body from the vertex of the cone.

Let X be the centre of the base of the cone and the plane face of the hemisphere. Then VX is an axis of symmetry of both the cone and the hemisphere and so will be an axis of symmetry of the composite body. The centre of mass G_1 of the cone lies on this line, the centre

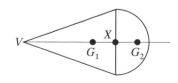

of mass G_2 of the hemisphere lies on this line and so the centre of mass of the composite body G will also lie on this line.

Recall that:

(i) the centre of mass of a uniform solid cone of height h is a distance $\dfrac{h}{4}$ from the base

(ii) the centre of mass of a uniform solid hemisphere of radius a is a distance $\dfrac{3a}{8}$ from the plane face.

So:
$$VG_1 = h - \frac{h}{4} = \frac{3h}{4}$$
$$VG_2 = VX + XG_2 = \left(h + \frac{3a}{8}\right)$$

Construct a table showing the masses of the separate bodies and the composite body, and the positions of the centres of mass as in chapter 6 of Book M1.

	Separate bodies		Composite body
	Cone	Hemisphere	
Mass	$\frac{1}{3}\pi a^2 h\rho$	$\frac{2}{3}\pi a^3 \rho$	$\frac{1}{3}\pi a^2 \rho(h+2a)$
Distance from V to centre of mass	$\dfrac{3h}{4}$	$\left(h+\dfrac{3a}{8}\right)$	$\bar{x}$

So:
$$M\bar{x} = \frac{1}{3}\pi a^2 \rho(h+2a)\bar{x} \tag{1}$$

And:
$$M\bar{x} = (\tfrac{1}{3}\pi a^2 h\rho)\frac{3h}{4} + (\tfrac{2}{3}\pi a^3 \rho)\left(h + \frac{3a}{8}\right)$$
$$= \tfrac{1}{4}\rho\pi a^2 h^2 + \tfrac{2}{3}\rho\pi a^3 h + \tfrac{1}{4}\rho\pi a^4$$
$$= \frac{\pi a^2 \rho}{12}\left[3h^2 + 8ah + 3a^2\right] \tag{2}$$

So from (1) and (2):
$$\bar{x} = \left(\frac{\pi a^2 \rho}{12}\right)\left(\frac{3}{\pi a^2 \rho}\right)\frac{[3h^2 + 8ah + 3a^2]}{(h+2a)}$$
$$= \frac{1}{4}\left(\frac{1}{h+2a}\right)[3h^2 + 8ah + 3a^2]$$

This is the distance of the centre of mass G of the composite body from the vertex of the cone.

Example 11

A uniform solid right circular cone has a base radius of 6 cm and a height of 12 cm. The top part of it, consisting of a small cone, base radius 3 cm and height 6 cm, is removed. Find the distance from the base of the large cone to the centre of mass of the remaining portion. (The remaining portion is called a **frustum**.)

A problem of this kind involving laminae was considered in chapter 6 of Book M1 (example 15). The method used here is exactly the same in that the complete body is considered to be formed from the final frustum and the small cone which is removed.

Let O be the centre of the base of the large cone. Since OV is an axis of symmetry of both cones it will also be an axis of symmetry of the frustum. The centre of mass G of the frustum therefore lies on OV.

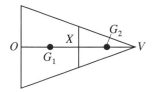

Let X be the point where OV meets the base of the small cone. Using the standard result for the centre of mass of a cone (see the table at the end of this chapter, or the formula book) then:

$$XG_2 = \tfrac{1}{4} \times 6 \, \text{cm}$$

and $\quad OG_2 = OX + XG_2 = 6 \, \text{cm} + \tfrac{3}{2} \text{cm} = 7\tfrac{1}{2} \text{cm}$

$$OG_1 = \tfrac{1}{4} \times 12 \text{ cm} = 3 \text{ cm}$$

Also: $\qquad$ mass of large cone $= \tfrac{1}{3}\pi(6)^2 12\rho = 144\pi\rho$

$\qquad\qquad$ mass of small cone $= \tfrac{1}{3}\pi(3)^2 6\rho = 18\pi\rho$

So: $\qquad\qquad$ mass of frustum $= 126\pi\rho$

	Separate bodies		Composite body
	Small cone	Frustum	Large cone
Mass	$18\pi\rho$	$126\pi\rho$	$144\pi\rho$
Ratio of masses	1	7	8
OG	$7\tfrac{1}{2}$ cm	$\bar{x}$	3 cm

So: $\qquad\qquad 1 \times 7\tfrac{1}{2} + 7\bar{x} = 8 \times 3$

or: $\qquad\qquad 7\bar{x} = 24 - 7\tfrac{1}{2} = 16\tfrac{1}{2}$

and $\qquad\qquad \bar{x} = 2\tfrac{5}{14} \, \text{cm}$

The distance from the base of the larger cone to the centre of mass of the frustum is $2\tfrac{5}{14}$ cm.

Exercise 6B

You are advised to draw a sketch in these exercises.

1 The vertices of the shaded triangular region shown in the figure are $O(0,0)$, $A(6,0)$ and $B(3,3)$. The region is rotated through $360°$ about the x-axis. By considering its symmetries write down the coordinates of the centre of mass of the solid of revolution formed.

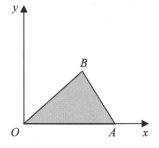

In questions 2, 3 and 4 use symmetry considerations to obtain the answer.

2 The finite region bounded by the curve with equation $y = x^2 - 2x$ and the x-axis is rotated completely about the x-axis. Find the coordinates of the centre of mass of the solid of revolution formed.

3 The finite region bounded by the x-axis and the curve with equation $y = \sin x$ between $x = 0$ and $x = \pi$ is rotated completely about the x-axis. Find the coordinates of the centre of mass of the solid of revolution formed.

4 The finite region bounded by the y-axis and the curve with equation $y^2 - y = x$ is rotated completely about the y-axis. Find the coordinates of the centre of mass of the solid formed.

5 Find the coordinates of the centre of mass of the solid formed when the finite region bounded by the curve with equation $y = x^2$, the line $x = 1$ and the x-axis is rotated completely about the x-axis.

6 Find the coordinates of the centre of mass of the solid formed when the finite region bounded by the curve with equation $y = x^3$, the line $x = 1$ and the x-axis is rotated completely about the x-axis.

7 Find the coordinates of the centre of mass of the solid formed when the finite region bounded by the curve with equation $y = (1 + x^2)$, the lines $x = 0$ and $x = 1$ and the x-axis is rotated completely about the x-axis.

8 Find the coordinates of the centre of mass of the solid formed when the finite region bounded by the curve with equation $y = \dfrac{1}{x}$, the lines $x = 1$ and $x = 2$ and the x-axis is rotated completely about the x-axis.

9 Find the coordinates of the centre of mass of the solid formed when the finite region bounded by the curve with equation $y = \dfrac{1}{x^2}$, the lines $x = 1$ and $x = 2$ and the x-axis is rotated completely about the x-axis.

10 Find the coordinates of the centre of mass of the solid formed when the finite region bounded by the curve with equation $y = e^x$, the lines $x = 0$ and $x = 1$ and the x-axis is rotated completely about the x-axis.

11 Find the coordinates of the centre of mass of the solid formed when the finite region bounded by the curve with equation $y = e^{-x}$, the lines $x = 0$ and $x = 1$ and the x-axis is rotated completely about the x-axis.

12 A uniform cone C has base radius r and height h. It is joined to a uniform solid right circular cylinder, of the same density, with radius r and length l, so that the plane base of C coincides with a plane face of the cylinder. Find the distance from the vertex of C of the centre of mass of the composite body.

13 The base radius of the uniform solid cone C_1 is r and the radius of the uniform solid right circular cylinder C_2 is also r. The cylinder and the cone are made of the same material. A composite body is formed by joining the two so that the plane face of the cone and an end of the cylinder coincide. The centre of mass of the composite body is at the centre of the base of the cone. Find the ratio of the height of the cone to the length of the cylinder.

14 A uniform solid cylinder is of radius 4 cm and height 6 cm. A cylindrical hole of radius 2 cm and depth 3 cm is bored centrally at one end E. Find the distance of the centre of mass of the remainder from the end E.

15 A uniform solid cone C is of height h and base radius r. A frustum is cut from the cone by a plane parallel to the base at a distance $\frac{1}{2}h$ from the vertex. Show that the centre of mass of the frustum is at a distance $\frac{11}{56}h$ from its larger plane end.

16 The radii of the plane faces of the frustum of a cone are 2 cm and 3 cm and the distance between the faces is 5 cm. A uniform paper collar is fitted to the curved surface of the frustum. Find the distance of the centre of mass of the paper collar from the centre of the larger plane face of the frustum.

6.3 Simple cases of equilibrium of rigid bodies

In this section we consider situations in 3 dimensions where all the forces act in a vertical plane through the centre of mass of the body. Such problems are essentially two dimensional and we have a system of coplanar forces. The results used in chapter 6 of Book M1 for laminae can be generalised to this situation.

■ A rigid body is in **equilibrium** if
  (i) the vector sum of the forces acting is zero, that is the sum of the components of the forces in any given direction is zero.
  (ii) the algebraic sum of the moments of the forces about any given point is zero.

Suspension of a body from a fixed point

In chapter 6 of Book M1 it was shown that for a suspended lamina to be in equilibrium its centre of mass G must be vertically below the point of suspension.

The same result holds for a rigid body.

Consider a rigid body suspended from a point A. The only forces acting on the body are the force at A and the weight of the body. For the body to be in equilibrium these forces must be equal in magnitude and act in opposite directions along the same vertical line. Hence:

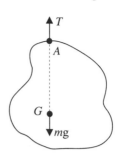

■ **a rigid body hangs in equilibrium with its centre of mass vertically below the point of suspension.**

Example 12

A uniform solid right circular cone has base radius r and height $8r$. The cone is suspended by a string attached to a point A on the rim of its base. Find the angle between the axis of the cone and the downward vertical in the equilibrium position.

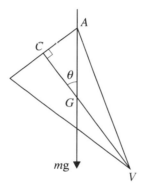

Using the standard formula $\dfrac{h}{4}$ for the distance of the centre of mass of a uniform solid cone from its base:

$$CG = \tfrac{1}{4}(8r) = 2r$$

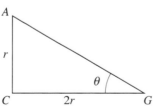

Let $\angle CGA$ be θ.

From the diagram: $\tan \theta = \dfrac{r}{2r} = \tfrac{1}{2}$

So: $\theta = 26.6°$

The angle between the axis of the cone and the downward vertical is $26.6°$.

As in the case of a lamina (see chapter 6 of Book M1) it is often possible to find the required angle without redrawing the solid. Simply add G to your original diagram, join AG, and label this line as the vertical.

Equilibrium of bodies on a horizontal plane

Consider a body in contact with a horizontal plane. The figure shows a cross-section through the body which contains the centre of mass G. Suppose there is an area of contact represented in this cross-section by the line AB.

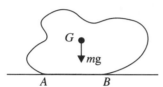

The weight of the body acts downwards through the centre of mass, and the only other vertical force acting on the body is the normal contact force or normal reaction R. If the body is to be in equilibrium in the position shown in the figure, the weight and R must be equal in magnitude and act in opposite directions along the same vertical line. Whether this is possible will depend on the position of G relative to the area of contact.

(i) If the line of action of the weight lies **inside** the area of contact as in the figure above, then the body is in equilibrium and the force diagram is:

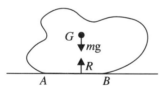

(ii) If the line of action of the weight lies **outside** the area of contact as shown here then it is not possible to have equilibrium as R must act somewhere in the area of contact. In this case the body will topple.

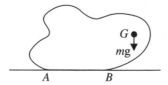

The intermediate case occurs when the line of action of the weight passes through either of the end points, A and B, of the area of contact. The body is then in equilibrium. This is the limiting case:

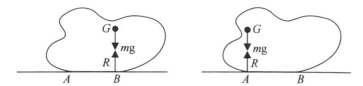

■ **For equilibrium the vertical through the centre of mass must fall in the area of contact.**

Example 13

A uniform square lamina $ABCD$ has side of length 4 cm. The point E lies on AD and is such that $ED = 3$ cm. The triangle EDC is removed. Show that if the lamina is placed in a vertical plane with AE on a rough horizontal plane it will topple.

A mass m is to be placed at the corner B. Find the smallest value of m as a fraction of M, the mass of $ABCD$, that will prevent toppling.

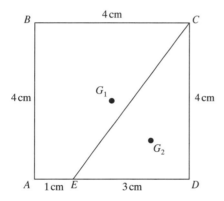

To calculate the distance of the centre of mass from AB proceed as in chapter 6 of Book M1.

The centre of mass of a square lies at the centre of the square and the centre of mass of a triangle is $\frac{2}{3}$ along a median from a vertex.

	Triangle	Remainder		Square
Mass	$\frac{1}{2} \times 3 \times 4\rho = 6\rho$	$16\rho - 6\rho = 10\rho$		$4 \times 4\rho = 16\rho$
Ratio of masses	3	5		8
Distance of centre of mass from AB	$4 - \frac{1}{3}(3) = 3$	$\bar{x}$		2
	(G_2)	(G)		(G_1)

where ρ is the mass per unit area.

Hence : $$3 \times 3 + 5 \times \bar{x} = 8 \times 2$$

$$5\bar{x} = 16 - 9 = 7$$

So: $$\bar{x} = 1.4\,\text{cm}$$

Since $AE = 1$ cm the line of action lies outside the area of contact and so the lamina will topple.

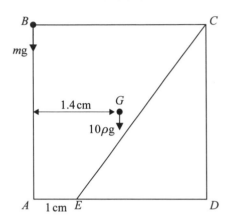

The body will turn about the point E when it topples. To prevent toppling the algebraic sum of the moments about E must be zero.

For lamina: moment about $E \curvearrowright (0.4)\,\text{cm} \times 10\rho g$

For mass: moment about $E \curvearrowleft (1)\,\text{cm} \times mg$

So: $$(0.4)10\rho = m$$

and: $$m = 4\rho$$

Since: $$M = 4 \times 4\rho$$

So: $$\frac{m}{M} = \frac{4\rho}{4 \times 4\rho} = \frac{1}{4}$$

Example 14

A uniform solid right circular cone has base radius 3 m and height 4 m. The top part of it, consisting of a small cone of base radius $1\frac{1}{2}$ m and height 2 m, is removed. Find the distance from the base of the larger cone to the centre of mass of the remaining portion and hence show that the frustum can rest in equilibrium with its curved surface in contact with a horizontal plane.

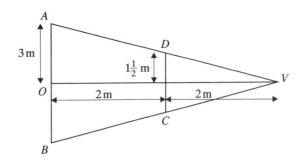

The calculation of the position of the centre of mass proceeds as before.

	Small cone	Frustum	Large cone
Mass	$\frac{1}{3}\pi\left(1\frac{1}{2}\right)^2 2\rho$ $= 1\frac{1}{2}\pi\rho$	$12\pi\rho - 1\frac{1}{2}\pi\rho$ $= 10\frac{1}{2}\pi\rho$	$\frac{1}{3}\pi(3)^2 4\rho$ $= 12\pi\rho$
Ratio of masses	1	7	8
Distance of centre of mass from AB	$2 + \frac{1}{4}(2) = 2\frac{1}{2}$	$\bar{x}$	$\frac{1}{4}(4) = 1$

Hence:
$$2\frac{1}{2} + 7\bar{x} = 8$$

or:
$$7\bar{x} = 8 - 2\frac{1}{2} = 5\frac{1}{2}$$

So:
$$\bar{x} = \frac{11}{14}\ \text{m}$$

Consider now the situation with the curved surface in contact with a horizontal plane.

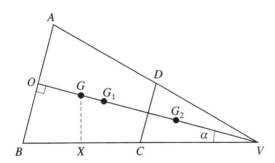

The triangle OBV is a right-angled triangle and since $OB = 3$ m and $OV = 4$ m, so by Pythagoras' theorem $BV = 5$ m. Also $CV = \frac{1}{2}BV = 2\frac{1}{2}$ m. The distance GV of G, the centre of mass of the frustum, from V is 4 m $- \frac{11}{14}$ m $= \frac{45}{14}$ m

If the line of action of the weight cuts BV at X then:

$$VX = \tfrac{45}{14} \cos \alpha$$

From $\triangle OBV$, $\cos \alpha = \tfrac{4}{5}$ and so:

$$VX = \tfrac{45}{14}\left(\tfrac{4}{5}\right) = \tfrac{18}{7} = 2\tfrac{4}{7}\,\text{m} = 2.57\,\text{m}$$

Since $VX > VC = 2\tfrac{1}{2}\,\text{m}$ the line of action of the weight lies inside the area of contact. The frustum can therefore rest in equilibrium with its curved surface in contact with the horizontal plane.

Equilibrium of bodies on an inclined plane

Consider a body in contact with an inclined plane. The figure shows a cross-section through the plane and the body which contains the centre of mass G and a line of greatest slope of the plane.

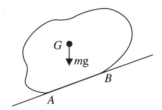

Suppose there is an area of contact represented in this cross-section by the line AB. The forces acting on the body are: the weight of the body; the normal contact force R, perpendicular to the plane; the frictional force F parallel to the plane. For equilibrium, these three forces must all act through the same point *and* the resultant of R and F must be equal in magnitude to the weight of the body and act vertically upwards.

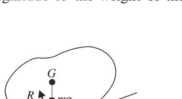

As in the previous discussion:

■ **if the line of action of the weight lies *inside* the area of contact the body is in equilibrium.**
If the line of action of the weight lies outside the area of contact then the body cannot rest in equilibrium and the body will topple.

Example 15

A solid uniform cylinder of radius r and length r has a solid uniform right circular cone, made from the same material, of base radius r and height r joined to it so that the base of the cone coincides with one end of the cylinder. The composite body is placed with its circular face on a rough plane which is inclined at an angle α to the horizontal. The plane is sufficiently rough to prevent sliding. Show that the body will topple over if $\tan \alpha > \frac{16}{11}$.

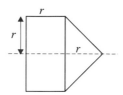

The calculation of the position of the centre of mass is shown in the usual tabular form:

	Cylinder	Cone	Composite body
Mass	$(\pi r^2) r\rho$	$\frac{1}{3}(\pi r^2) r\rho$	$\frac{4}{3}\pi r^3 \rho$
Ratio of masses	3	1	4
Distance of G from circular end	$\frac{r}{2}$	$r + \frac{r}{4} = \frac{5r}{4}$	$\bar{x}$

where ρ is mass per unit volume.

So:
$$3 \times \frac{r}{2} + 1 \times \frac{5r}{4} = 4 \times \bar{x}$$

$$\frac{11r}{4} = 4\bar{x}$$

And:
$$\bar{x} = \frac{11}{16}r$$

The centre of mass lies on the axis of symmetry.

The limiting case is shown in the figure:

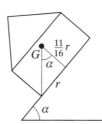

$$\tan \alpha = \frac{r}{\left(\frac{11}{16}\right)r} = \frac{16}{11}$$

So for the limiting case:

The body will therefore topple if α exceeds this value, that is if $\tan \alpha > \frac{16}{11}$

Exercise 6C

1 A uniform solid hemisphere is suspended by a string attached to a point on the rim of its base. Find the angle between the horizontal and the plane face of the hemisphere when it is in equilibrium.

2 A toy is constructed of thin uniform sheet metal and consists of a right circular cylinder with open ends of length l and radius r $(r < l)$ covered by a hemispherical cap of radius r. Show that the centre of mass of the toy is at a distance $\frac{1}{2}(l + r)$ from the open end.

   The toy is suspended from a point on the rim of the hemispherical cap. Show that, when the toy hangs in equilibrium, the axis of symmetry makes an angle α to the horizontal where:

   $$\tan \alpha = \frac{l - r}{2r}$$

3 A square lamina $ABCD$ of side $4\,\text{cm}$ is made of uniform thin material. A semicircular piece, with AB as diameter, is removed. Show that the centre of mass of the remaining shape is at a distance:

   $$\frac{40}{3(8 - \pi)}\ \text{cm}$$

   from AB.

   The remaining shape is suspended by a string attached at C and hangs in equilibrium. Show that the angle θ between CD and the downward vertical is given by:

   $$\tan \theta = \frac{2(14 - 3\pi)}{3(8 - \pi)}$$

4 A uniform right circular cone of base radius a and height $3a$ rests in equilibrium with its base in contact with a rough inclined plane. The angle of inclination of the plane is then increased steadily. Assuming the plane is sufficiently rough to prevent slipping, find the angle of inclination of the plane to the horizontal when the cone is about to topple.

5 A uniform thin hemispherical bowl rests in equilibrium with its axis of symmetry horizontal and its curved surface on a rough inclined plane. Find the angle between the plane and the horizontal.

6 A uniform solid consists of a hemisphere of radius r and a right circular cone of base radius r and height h, fixed together so that their plane faces coincide. The solid can rest in equilibrium with any point of the curved surface of the hemisphere in contact with a horizontal plane. Find h in terms of r.

7 A uniform solid spinning top has the shape of an inverted right circular cone of radius $3r$ and height $4r$ surmounted by a right circular cylinder of base radius $3r$ and height $6r$. Find the position of the centre of mass of the spinning top and hence show that if it is placed with the curved surface of the cone on a horizontal plane the top will topple.

6.4 Simple systems involving three forces

The conditions for a body to be in equilibrium were stated in section 6.3. The following procedure is suggested for problems which involve several forces:

(1) **Draw** a clearly labelled diagram showing all the forces acting on the body.

(2) **Resolve** the forces in each of two perpendicular directions and equate the sum of the resolved parts to zero. (Usually the best directions in which to resolve are horizontally and vertically or parallel and perpendicular to an inclined plane.)

(3) **Take moments** about any convenient point and equate the algebraic sum of the moments to zero. (Careful choice of the point used here may greatly simplify the solution in a particular problem.)

Example 16

A uniform rod AB of length $2a$ and mass m is smoothly hinged at A. It is maintained in equilibrium by a horizontal force of magnitude P acting at B. Given that the rod is inclined at $30°$ to the horizontal with B below A find:

(a) the value of P,

(b) the magnitude and direction of the reaction at the hinge.

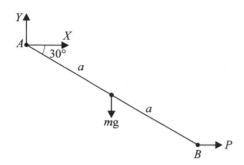

The reaction at the hinge is assumed to have horizontal and vertical components X and Y respectively.

Resolving horizontally: $\qquad \rightarrow X + P = 0 \qquad$ (1)

Resolving vertically: $\qquad \uparrow Y - mg = 0 \qquad$ (2)

As the forces X and Y act through A, taking moments about A will produce an equation not involving X and Y.

(a) Taking moments about $A \curvearrowleft$

$$P \times 2a \sin 30° - mga \cos 30° = 0 \qquad (3)$$

Using $\sin 30° = \frac{1}{2}$ and $\cos 30° = \frac{\sqrt{3}}{2}$ equation (3) gives:

$$Pa - \frac{\sqrt{3}}{2} mga = 0$$

So: $\qquad\qquad\qquad P = \frac{\sqrt{3}}{2} mg \qquad (4)$

(b) From equation (2): $\qquad Y = mg$

Using equations (1) and (4):

$$X = -P = -\frac{\sqrt{3}}{2} mg$$

The reaction at the hinge is:

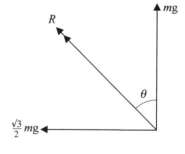

$$R = \sqrt{\left((mg)^2 + \left(\frac{\sqrt{3}}{2} mg \right)^2 \right)}$$
$$= mg \sqrt{\left(1 + \frac{3}{4} \right)} = \frac{\sqrt{7}}{2} mg$$

And : $\qquad\qquad \tan \theta = \dfrac{\frac{\sqrt{3}}{2} mg}{mg} = \frac{\sqrt{3}}{2}$

so : $\qquad\qquad \theta = 40.9°$

Hence the reaction at the hinge has magnitude $\frac{\sqrt{7}}{2} mg$ and makes an angle of $40.9°$ with the vertical.

Limiting equilibrium

When one of the forces acting on a body is a frictional force and the body is in equilibrium it is necessary to be clear whether the body is in **limiting equilibrium**. (Limiting equilibrium is discussed in chapter 4 of Book M1.) It is to be stressed that **only in the case of limiting equilibrium**, when motion is on the point of taking place, **does the frictional force F have its maximum value μR** (where μ is the coefficient of friction).

Example 17

A uniform rod AB, of mass $10\,\text{kg}$, rests with its lower end A on a rough horizontal floor. The coefficient of friction between the rod and the floor is μ. One end of a string is attached to end B of the rod and the other end is fixed so that the string is perpendicular to the rod. The rod rests in equilibrium at an angle of $30°$ to the horizontal. Find:

(a) the tension T in the string,

(b) the magnitudes of R the normal contact force and F the frictional force.

(c) Determine the least possible value of μ for equilibrium to be possible.

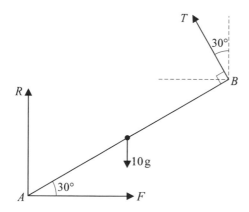

The diagram includes all the given information together with all the forces acting. Since A would move to the left if slipping occurred, the frictional force must act towards the right.

Resolving horizontally: $\qquad\qquad \rightarrow F - T\sin 30° = 0 \qquad (1)$

Resolving vertically: $\qquad\qquad \uparrow R + T\cos 30° - 10g = 0 \qquad (2)$

Let the rod have length $2a$.

Since both R and F act through the point A the simplest possible moment equation is obtained by taking moments about A.

Taking moments about A: $\qquad \curvearrowleft T2a - 10ga\cos 30° = 0 \qquad$ (3)

From (3): $\qquad\qquad 2T - 10g\frac{\sqrt{3}}{2} = 0$ using $\cos 30° = \frac{\sqrt{3}}{2}$

(a) So: $\qquad\qquad\qquad T = \frac{5\sqrt{3}g}{2} = 42.4\,\text{N} \qquad\qquad$ (4)

(b) Using this result in (1):

$$F = T\sin 30° = T(\tfrac{1}{2}) \text{ using } \sin 30° = \tfrac{1}{2}$$
$$= 21.2\,\text{N}$$

Using (4) in (2):

$$R = 10g - T\cos 30°$$
$$= 10g - \frac{T\sqrt{3}}{2} = 10g - \frac{5 \times 3}{4}g$$
$$= 61.3\,\text{N}$$

(c) Since $\mu = \dfrac{F_{max}}{R}$ the least value of μ necessary is $\dfrac{21.2}{61.3} = 0.346$ and the rod would then be in limiting equilibrium.

Problems involving ladders

The situation of a ladder resting against a wall with the foot of the ladder on the ground occurs frequently.

It may be modelled by a uniform rod in contact with a vertical wall and resting on horizontal ground. There are some circumstances where the frictional forces on the rod due to the wall or the floor are so small that they may, as a first model, be ignored; that is the contacts may be regarded as 'smooth'. In other cases the frictional forces cannot be neglected and so the contacts will be 'rough'.

When the rod is in contact with a **smooth** surface there will only be a normal contact force R, so $F = 0$. When the rod is in contact with a **rough** surface the frictional force F acts parallel to the surfaces in contact in such a direction as to oppose subsequent motion.

Example 18
A uniform ladder of mass 10 kg and length 6 m rests against a smooth vertical wall with its lower end on rough ground. The ladder rests in equilibrium at an angle of 60° to the horizontal. Find:

(a) the magnitude of the normal contact force S at the wall,

(b) the magnitude of the normal contact force R at the ground and the frictional force at the ground.

(c) Obtain the least possible value of μ.

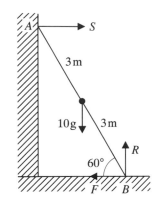

The ladder is modelled by a uniform rod. The contact at A is perfectly smooth and the contact at B is rough.

Resolving horizontally: $\rightarrow S - F = 0$ (1)

Resolving vertically: $\uparrow R - 10\,g = 0$ (2)

Moments about B: $\curvearrowright S \times (6\sin 60°) - 10\,g \times (3\cos 60°) = 0$ (3)

Using $\sin 60° = \frac{\sqrt{3}}{2}$ and $\cos 60° = \frac{1}{2}$:

$$S3\sqrt{3} - 10\,g \times \tfrac{3}{2} = 0$$

(a) So: $$S = \frac{10\,g}{2\sqrt{3}} = 28.3\,\text{N}$$

(b) From (1): $F = S = 28.3\,\text{N}$

From (2): $R = 10\,g = 98\,\text{N}$

(c) Since: $$\mu = \frac{F_{\text{max}}}{R}$$

and: $$\frac{F}{R} = \frac{28.3}{98} = 0.289$$

μ must be at least 0.289.

Climbing a ladder

It is very important to know whether or not it is safe to climb to the top of a ladder. The answer will depend on the magnitude of the frictional force acting on the ladder. This in turn depends on the roughness of the ground on which the ladder rests. If the ladder is found to be in limiting equilibrium when a person is only part way up the ladder then any further ascent will cause the ladder to slip.

Such a problem is dealt with in the next example, where the ladder is modelled by a uniform rod and the person by a particle at the point where he stands on the ladder.

Example 19

A uniform ladder of mass 30 kg and length 10 m rests against a smooth vertical wall with its lower end on rough ground. The coefficient of friction between the ground and the ladder is 0.3. The ladder is inclined at an angle θ to the horizontal where $\tan \theta = 2$. Find how far a boy of mass 30 kg can ascend the ladder without it slipping.

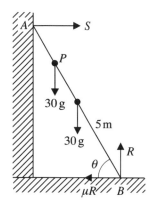

Let the boy be at the point P, where $BP = l$ m, when the ladder is about to slip.

Since the ladder is about to slip the frictional force is μR.

Resolving horizontally: $\rightarrow$ $S - \mu R = 0$ (1)

Resolving vertically: $\uparrow R - 30\,g - 30\,g = 0$ (2)

Taking moments about B:

$$\curvearrowright S \times 10 \sin \theta - 30\,g \times (5 \cos \theta) - 30\,g \times (l \cos \theta) = 0 \qquad (3)$$

From equation (2): $R = 60\,g$

Substituting into (1) with $\mu = 0.3$:

$$S = (0.3)60\,g = 18\,g \qquad (4)$$

Using (4) in equation (3) gives:

$$18 \times 10\,g \sin \theta - 30 \times 5\,g \cos \theta - 30g \times l \cos \theta = 0$$

So: $6 \tan \theta - 5 = l$

Using $\tan \theta = 2$ gives $l = 7$.

So the boy can ascend 7 m up the ladder before it begins to slip.

Example 20

A uniform rod AB of mass 20 kg and length 8 m rests with the end A on rough horizontal ground. The rod rests against a smooth peg C, where $AC = (4 + x)$ m. The rod is in limiting equilibrium inclined at an angle of 45°. Given that the coefficient of friction between the rod and the ground is $\frac{1}{2}$, find:

(a) the value of x,

(b) the magnitude of the reaction at C.

The force acting at C is perpendicular to the rod since the peg is smooth. Since the rod is in limiting equilibrium the frictional force at the lower end of the rod is μR, where R is the normal contact force acting there.

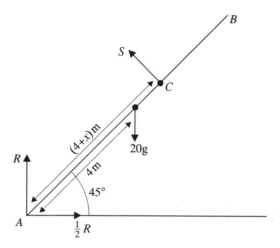

Resolving horizontally: $\quad\rightarrow\quad \frac{1}{2}R - S\sin 45° = 0 \qquad (1)$

Resolving vertically: $\quad\uparrow\ R + S\cos 45° - 20\,g = 0 \qquad (2)$

Taking moments about A: $\quad\curvearrowleft S(4 + x) - 20\,g \times 4\cos 45° = 0 \quad (3)$

From (1): $\quad \frac{1}{2}R - S\frac{\sqrt{2}}{2} = 0 \quad$ since $\sin 45° = \frac{\sqrt{2}}{2}$

So: $\qquad\qquad\qquad S = \dfrac{R}{\sqrt{2}} \qquad\qquad (4)$

From (2): $\qquad R + S\frac{\sqrt{2}}{2} - 20\,g = 0$ since $\cos 45° = \frac{\sqrt{2}}{2}$

Substituting for S from (4):

$$R + \frac{R}{2} - 20\,g = 0$$

So:

$$\frac{3}{2} R = 20\,g$$

and:

$$R = \frac{40\,g}{3}$$

and using (4):

$$S = \frac{40\,g}{3\sqrt{2}} = 92.4\,\text{N} \qquad (5)$$

From (3): $S(4 + x) = 20\,g \times \frac{4\sqrt{2}}{2}$ since $\cos 45° = \frac{\sqrt{2}}{2}$

Using (5):

$$\frac{40\,g}{3\sqrt{2}}(4 + x) = 20\,g \times \frac{4\sqrt{2}}{2}$$

So:

$$4 + x = 6$$

and:

$$x = 2$$

(a) The value of x is 2.

(b) The reaction of C is of magnitude 92.4 N.

Exercise 6D

1 A uniform ladder rests in limiting equilibrium with its top against a smooth vertical wall and its base on a rough horizontal floor. The coefficient of friction between the ladder and the floor is μ. Given that the ladder makes an angle θ with the floor show that:

$$2\mu \tan \theta = 1$$

2 A uniform ladder of mass 30 kg and length 10 m rests with one end on a smooth horizontal floor and the other end against a smooth vertical wall. The ladder is held in this position by a light inextensible string of length 5 m which has one end attached to the bottom of the ladder and the other end fastened to a point at the base of the wall vertically below the top of the ladder. Find the tension in the string.

3 A uniform ladder AB of mass 30 kg rests with its top A against a smooth vertical wall and its base B on rough horizontal ground. A mass of 30 kg is attached to the ladder at A. The coefficient of friction between the ladder and the ground is $\frac{3}{4}$. Given that the ladder is just about to slip, find the inclination of the ladder to the horizontal.

4 A uniform ladder of mass 30 kg and length 10 m rests with its top against a smooth vertical wall and its base on rough horizontal ground. The ladder rests in equilibrium at $60°$ to the horizontal with a man of mass 90 kg standing on the ladder at a point 7.5 m from its base.

(a) Find the magnitude of the normal contact force and the frictional force exerted on the ladder by the ground.

(b) Find the minimum value of the coefficient of friction between the ladder and the ground that would enable the man to climb to the top of the ladder.

5 A uniform ladder rests with one end on rough horizontal ground and the other end against a rough vertical wall. The coefficient of friction between the ground and the ladder is $\frac{3}{5}$ and the coefficient of friction between the wall and the ladder is $\frac{1}{3}$. The ladder is on the point of slipping when it makes an angle α with the horizontal. Find $\tan \alpha$.

6 A uniform rod AB of mass 40 kg and length 10 m rests with the end A on rough horizontal ground. The rod rests against a smooth peg C where $AC = 8$ m. The rod is in limiting equilibrium at an angle of $\alpha°$ to the horizontal. Given that the reaction at C is of magntiude 196 N find:
(a) the value of α,
(b) the coefficient of friction between the rod and the ground.

7 A uniform ladder of mass 30 kg is placed with its base on rough horizontal ground. The coefficient of friction between the ladder and the ground is $\frac{1}{4}$. The upper end of the ladder rests against a smooth vertical wall, the ladder making an angle of $60°$ to the horizontal. Find the magnitude of the minimum horizontal force that must be applied to the base of the ladder to prevent slipping.

8 A uniform rod of mass 20 kg and length 20 m rests in limiting equilibrium with the upper end against a smooth vertical wall and the lower end on a rough horizontal floor.

(a) Given that the coefficient of friction is $\frac{1}{2}$, find the angle made by the ladder to the horizontal.

(b) Describe a 'physical situation' for which the above could be used as a model and state what assumptions would be made in using this model.

(c) How could this model be refined to take into account other attributes of the 'physical situation'?

9 A uniform ladder of mass M rests in limiting equilibrium with one end on rough horizontal ground and the other end against a rough vertical wall. The coefficient of friction between the ladder and the ground is μ and the coefficient of friction between the ladder and the wall is μ'. Given that the ladder makes an angle α with the horizontal, show that:

$$\tan \alpha = \frac{1 - \mu\mu'}{2\mu}$$

SUMMARY OF KEY POINTS

1

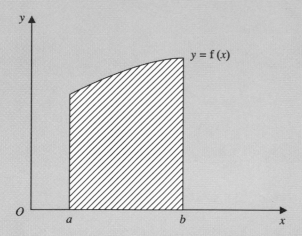

The coordinates $(\bar{x}, \bar{y})$ of the centre of mass of the shaded region are given by:

$$\left[\int_a^b f(x)dx\right]\bar{x} = \int_a^b x f(x)dx$$

$$\left[\int_a^b f(x)dx\right]\bar{y} = \tfrac{1}{2}\int_a^b [f(x)]^2 dx$$

2 The centre of mass of a solid body is the point at which the weight acts.

The weight of a uniform solid body is evenly distributed throughout its volume.

The centre of mass of a uniform solid body must lie on any **axis** of symmetry.

The centre of mass of a uniform solid body must lie on any **plane** of symmetry.

The centre of mass of a system of n masses $m_1, m_2, ..., m_n$ at points with coordinates $(x_1, y_1, z_1), (x_2, y_2, z_2), ..., (x_n, y_n, z_n)$ has coordinates $(\bar{x}, \bar{y}, \bar{z})$ where

$$\bar{x} = \frac{\sum m_i x_i}{\sum m_i}$$

$$\bar{y} = \frac{\sum m_i y_i}{\sum m_i}$$

$$\bar{z} = \frac{\sum m_i z_i}{\sum m_i}$$

3

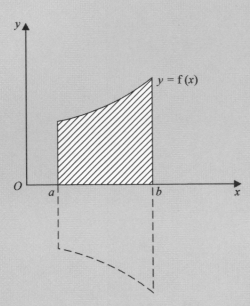

The x coordinate of the centre of mass, $\bar{x}$, of the solid of revolution generated when the shaded region is rotated through $360°$ about the x-axis is given by:

$$M\bar{x} = \left[\int_a^b \rho\pi y^2 \mathrm{d}x \right] \bar{x} = \int_a^b \rho\pi y^2 x\,\mathrm{d}x$$

4 Standard results for uniform bodies

Body	Centre of mass
Solid hemisphere, radius r	$\frac{3}{8}r$ from centre
Hemispherical shell, radius r	$\frac{1}{2}r$ from centre
Circular arc, radius r, angle at centre 2α	$\dfrac{r\sin\alpha}{\alpha}$ from centre
Wire in form of a semicircle, radius r	$\dfrac{2r}{\pi}$ from centre
Sector of circle, radius r, angle at centre 2α	$\dfrac{2r\sin\alpha}{3\alpha}$ from centre
Semicircular lamina, radius r	$\dfrac{4r}{3\pi}$ from centre
Solid right circular cone, height h	$\dfrac{3h}{4}$ from vertex

5 A rigid body is in **equilibrium** if:
   (i) The vector sum of the forces acting is zero, that is the sum of the components of the forces in any given direction is zero.
   (ii) the algebraic sum of the moments of the forces about any given point is zero.

6 A rigid body hangs in equilibrium with its centre of mass vertically below the point of suspension.
   For a body in contact with a horizontal or inclined plane to be in equilibrium, the line of action of the weight, through the centre of mass, must lie inside the area of contact.

7 Only in the case of limiting equilibrium, when motion is on the point of taking place, does the frictional force F have its maximum value μR.

Review exercise

2

Whenever a numerical value of g is required take $g = 9.8 \, \text{m s}^{-2}$.

1 A child of mass 20 kg is seated on a swing, the ropes of which are 2 m long. She is pulled back by her father through an angle of 35° and pushed forwards with a speed of $2 \, \text{m s}^{-1}$. It is required to find her speed at the lowest point and the height she reaches, above this lowest point, on the other side.

It is suggested that the problem can be modelled by a particle of mass 20 kg attached to the end of a light inextensible string.

(a) Solve this model to find the required speed and distance.

(b) State the assumptions made using this model.

(c) Suggest ways in which this model can be refined.

2 A particle of mass m is suspended from a fixed point O by a light inextensible string of length a. When the string is hanging vertically the particle is projected horizontally with speed V. Find an expression for the tension in the string when it is inclined at an angle θ to the downward vertical.

Given that $V^2 = \frac{7}{2}ag$, show that the string becomes slack when $\theta = \frac{2}{3}\pi$ and find the greatest height reached by the particle above the horizontal plane through the point of projection.

[L]

3 A smooth circular hoop, centre O and radius a, is fixed with its plane vertical. A small smooth bead P, of mass $2m$, is threaded on the hoop. The bead is released from rest at a point on the same horizontal level as O. Show that if θ, $0 \leqslant \theta < \pi/2$, is the angle OP has turned through in time t

$$a\left(\frac{d\theta}{dt}\right)^2 = 2g\sin\theta.$$

When the bead reaches the lowest point A of the hoop it collides and coalesces with a small smooth bead of mass m, which is also threaded on the hoop and is at rest. Find the height, above A, to which the beads rise before coming to instantaneous rest.

Find also the magnitude of the force exerted by the beads on the hoop when the beads first return to A. [L]

4 A particle of mass m is attached to one end A of a light inelastic string of length l. The other end of the string is attached to a fixed point B and the particle hangs in equilibrium under gravity. A horizontal impulse of magnitude J is applied to the particle.

(a) Find the least possible value of J so that the particle can subsequently describe a complete vertical circle.

(b) Given that $J = m\sqrt{(3lg)}$, find the tension in the string when it first becomes horizontal.

Find also, in case (b), the height above the level of B to which the particle rises before the string first becomes slack. [L]

5 A particle P is projected horizontally with speed u, where $u^2 < ag$, from the highest point A of a fixed smooth sphere of radius a and whose centre is at O. The particle slides on the outer surface of the sphere. Show that it leaves the sphere when PO makes an angle $\arccos\left(\dfrac{2}{3} + \dfrac{u^2}{3ag}\right)$ with the vertical.

At this instant the speed is three times the initial speed. Show that $u = \sqrt{(ag/13)}$. [L]

6 A small bead of mass m slides on a smooth circular hoop, with centre O and radius a, which is fixed in a vertical plane. The bead is projected with speed u, where $2ag < u^2 < 4ag$, from A, the lowest point of the hoop, and at a subsequent time t the bead is at a point P, where $\angle POA = \theta$. Find in terms of u, a, g and θ, an expression for $\left(\dfrac{\mathrm{d}\theta}{\mathrm{d}t}\right)^2$ and deduce that R, the magnitude of the force exerted by the hoop on the bead, is given by

$$aR = \mid mu^2 - mga(2 - 3\cos\theta)\mid.$$

The bead is instantaneously at rest at the point B and the value of R is instantaneously zero at the point C. Show that the ratio of the vertical heights of B and C above the horizontal diameter of the hoop is $3:2$. [L]

7   A particle of mass m hangs at rest at one end of a light inextensible string of length l, the other end being fixed at the point O. The particle is then projected horizontally with speed u, where $u > \sqrt{(2gl)}$. Find the magnitude of the tension in the string when the string makes an acute angle α with the downward vertical through O.

When the string first becomes horizontal it comes into contact with a thin horizontal bar which is perpendicular to the plane of the motion and is at a distance $2l/3$ from O. Show that the particle describes complete vertical circles about the bar, provided that $u^2 \geqslant 3gl$. [L]

8   A smooth sphere, with centre O and radius a, is fixed to the upper surface of a table. A smooth particle P is placed on the surface of the sphere, at a point A, where OA makes an angle β with the upward vertical $\left(0 < \beta < \frac{\pi}{2}\right)$. The particle is released from rest. Show that when OP makes an angle θ with the upward vertical, and the particle is still in contact with the sphere,

$$a\dot{\theta}^2 = 2g(\cos\beta - \cos\theta).$$

Determine the value of θ when the particle leaves the sphere. Given that $\cos\beta = \frac{3}{4}$, show that the particle hits the table at a distance $(\frac{3}{4}\sqrt{3})a$ from the vertical diameter. [L]

9   A particle P of mass $2m$ is fastened to one end of a light elastic string of natural length l and modulus of elasticity $4mg$. The other end of the string is fastened to a fixed point A. The particle hangs at rest at a point B vertically below A.

(a) Show that the extension of the string is $\frac{l}{2}$.

The particle is projected vertically downwards from B with speed $\sqrt{\left(\frac{gl}{2}\right)}$. When the particle is at a point C below B, where $BC = x$, its speed is v. Find

(b) the increase in the elastic potential energy in the string as P moves from B to C, in terms of m, g, l and x,

(c) the loss of kinetic energy plus the loss of gravitational potential energy of P as it moves from B to C, in terms of m, g, l, x and v.

Hence show that the particle first comes to instantaneous rest at the point D where $BD = \dfrac{l}{2}$.

(d) Find, in terms of g and l, the speed of P when it is at the mid-point of BD. [L]

10 Show that the work done in stretching a light elastic string by a length x from its natural length a is $\frac{1}{2}Tx$, where T is. the final tension in the string.

One end A of a light elastic string AB of natural length a is attached to a fixed point at a height $a + b$ above a horizontal floor, where $b > 0$. To the other end B is attached a small spherical ball of mass m. The modulus of elasticity of the string is $\dfrac{mga}{b}$.

(a) Show that in equilibrium under gravity the ball is just in contact with the floor.

The ball is released from rest at A.

(b) Show that it strikes the floor with speed v, given by

$$v^2 = (2a + b)g.$$

When the ball strikes the floor the elastic string breaks and the ball rebounds to a height a.

(c) Find, in terms of a and b, the coefficient of restitution between the ball and the floor. [L]

11 Using Hooke's law, show by integration that the work done in stretching a light elastic string, of natural length l and modulus of elasticity λ, from length l to length $l + e$ is $\dfrac{\lambda e^2}{2l}$.

A light elastic string, of natural length 2 metres and modulus of elasticity $4mg$ newtons, has one end attached to a fixed point A. A particle P, of mass m kilograms, is attached to the other end of the string.

(a) Show that, when P hangs in equilibrium, the length of AP is $2\frac{1}{2}$ m.

The particle P is released from rest at A and falls vertically. Use the work–energy principle to calculate

(b) the speed, to 3 significant figures, of P at distance $2\frac{1}{2}$m below A,

(c) the greatest length of the string. [L]

12 A light elastic string AB has natural length l m and modulus of elasticity λ N. The end A is attached to a fixed point. When B is held at rest at a distance 0.6 m from A, the tension in the string is 3 N.

(a) Find an equation relating λ and l.

When B is held at rest at a distance 0.7 m from A, the elastic energy in the string is 0.6 J.

(b) Show that $l = 0.5$ and find the value of λ.

A particle, of mass 0.2 kg, is attached to the string at B with the end A still fixed. The particle B moves at constant speed in a circular path whose centre is vertically below A, so that the string AB traces out the curved surface of a cone. Given that the tension in the string is 3.92 N, calculate

(c) the angle between AB and the downward vertical at A,

(d) length of AB in this case, giving your answer in m to 2 decimal places.

(c) the speed of B, giving your answer in m s^{-1} to 2 decimal places. [L]

13 A light elastic string of natural length $30a$ is attached at its ends to points A and B which are in the same horizontal line and at a distance $30a$ apart. A particle of mass m is attached to the mid-point of the string and hangs in equilibrium at a distance $20a$ below AB.

(a) Show that the modulus of the string is $\frac{15}{16}mg$.

The particle is released from rest at O, the mid-point of the line AB. The particle has speed v when it is a distance x below O.

(b) Show that

$$v^2 = 2gx - \frac{1}{8}\left(\frac{g}{a}\right)\left[(225a^2 + x^2)^{\frac{1}{2}} - 15a\right]^2.$$

(c) Hence, or otherwise, show that the particle never reaches a distance of $37a$ below O. [L]

14 A light elastic string of natural length 5 m and modulus 4 N has one end fixed to a point O on level ground. To the other end of the string is attached a ball of mass 0.5 kg, which is projected vertically upwards from O with speed u m s^{-1}.

(a) Find u, given that the ball first comes to instantaneous rest when it reaches a height of 10 m above the ground.

Given that the coefficient of restitution between the ball and the ground is 0.6

(b) show that, after the first bounce at O, the string does not become taut. [L]

15 Particle A, of mass $3m$, moving with speed u on a horizontal plane, strikes directly a particle B, of mass m, which is at rest on the plane. The coefficient of restitution between A and B is e.

(a) Find, in terms of e and u, the speeds of A and B immediately after their collision.

Given that the magnitude of the impulse exerted by A on B is $\dfrac{5mu}{4}$,

(b) show that $e = \frac{2}{3}$.

Particle B then slides a distance $\dfrac{3u^2}{g}$ and strikes directly particle C, of mass λm, which is at rest on the plane. The coefficient of restitution between B and C is $\frac{2}{3}$ and the coefficient of friction between B and the plane is $\frac{1}{4}$. By considering the work done against friction, or otherwise,

(c) show that B strikes C with speed $\frac{1}{4}u$.

Given that B comes to rest immediately after striking C,

(d) find the value of λ. [L]

16 Two smooth spheres, A and B, of equal radii and masses m and $4m$ respectively, are moving on the surface of a smooth horizontal table. The sphere A, moving with speed u, overtakes and strikes directly the sphere B which is moving in the same direction with speed λu, where $0 < \lambda < 1$. The sphere A is brought to rest by the impact. Show that e, the coefficient of restitution between A and B, is given by

$$e = \frac{4\lambda + 1}{4(1 - \lambda)}.$$

Deduce that $\lambda \leqslant \frac{3}{8}$.

Given further that 25% of the total initial kinetic energy is lost in the collision between A and B, prove that

$$\lambda = (\sqrt{6} - 2)/2. \qquad [L]$$

17 A small smooth sphere S, of mass $0.2\,\text{kg}$, moving with speed $5\,\text{m s}^{-1}$ overtakes and collides with a second small smooth sphere T, of mass $0.3\,\text{kg}$, which is moving with speed $2\,\text{m s}^{-1}$ in the same direction as S. The coefficient of restitution between the spheres is 2/3. Calculate

(a) the speed of each sphere immediately after the impact,

(b) the magnitude of the impulse received by each sphere on impact.

18 A particle of mass m moving in a straight line with speed u receives an impulse of magnitude I in the direction of its motion. Show that the increase in kinetic energy is given by

$$I(I + 2mu)/(2m) \qquad [L]$$

19 Three particles A, B and C have masses m, $3m$ and λm respectively. The particles lie at rest on a smooth horizontal plane in a straight line with B between A and C. Particle A is given a horizontal impulse, of magnitude J, and collides directly with B. After this collision A is at rest and B moves towards C with speed u. The coefficient of restitution at each impact is e.

(a) Find J in terms of m and u.

(b) Show that $e = \frac{1}{3}$.

(c) Find in terms of m and u, the loss in kinetic energy in the collision between A and B.

Particle B, moving with speed u, collides directly with particle C.

(d) Find, in terms of λ and u, the speeds of B and C after their collision.

(e) Show that A and B will have a second collision provided that $\lambda > 9$.

(f) Given that $\lambda = 6$, find, in terms of m and u, the magnitude of the impulse on B in the collision between B and C.   [L]

20 The particles A and B of mass $3m$ and $2m$ respectively are moving due east in the same straight line on a smooth

horizontal plane. They collide directly. The coefficient of restitution between the particles is $\frac{3}{7}$. After the collision, A and B continue to move due east in the same line with speed $6u$ and $9u$ respectively.

(a) Calculate the speeds of A and B before the collision.

(b) Calculate the magnitude of the impulse exerted by A on B due to the collision.

(c) Show that the kinetic energy lost in the collision is $24mu^2$. [L]

21 A sphere of mass m moving along a smooth horizontal table with speed V collides directly with a stationary sphere of the same radius and of mass $2m$. Obtain expressions, in terms of V and the coefficient of restitution e, for the speeds of the two spheres after impact.

Half of the kinetic energy is lost in the impact. Find the value of e. [L]

22 The coefficient of restitution between two particles A and B is e, where $0 < e < 1$. The masses of A and B are m and em respectively. The particles are moving with constant speeds u and eu in the same horizontal line and in the same direction, as shown in the diagram, and they collide.

(a) Show that after the collision the speed of A is $u(1 - e + e^2)$ and that the speed of B is independent of e.

(b) Find the value of e for which the speed of A after the collision is least and deduce that, in this case, the total loss of kinetic energy due to the collision is $\frac{1}{32}mu^2$.

(c) Find the possible values of e for which the impulse of the force exerted by B on A due to the collision has magnitude $\frac{6}{25}mu$. [L]

23 A light inextensible string of length $4a$ has a particle A, of mass m, attached at one end and a particle B, of mass m, attached at the other end. The string passes through a small smooth ring

which is fixed at a point O at a distance $3a$ above a horizontal table. The system is hanging in equilibrium with $OB = 2a$ when a smooth bead of mass $2m$, which is threaded on the string between O and B, is released from rest at O. The bead falls under gravity until it collides with and adheres to the particle B to form a composite particle C.

(a) Given that the string remains taut, show that the speed of C immediately after the collision is $\sqrt{(ga)}$.

(b) Find the speed of C immediately before it reaches the table. C is now fixed to the table at the point directly below O and the particle A is set in motion so that it moves in a horizontal circle with constant speed $2\sqrt{(ag)}$.

(c) Find the cosine of the angle AOC.

(d) Find, in terms of m and g, the tension in the string.   [L]

24 A light elastic string of natural length $5\,\text{m}$ and modulus $4\,\text{N}$ has one end fixed to a point O on level ground. To the other end of the string is attached a ball of mass $0.5\,\text{kg}$, which is projected vertically upwards from O with speed $u\,\text{m}\,\text{s}^{-1}$.

(a) Find u, given that the ball first comes to instantaneous rest when it reaches a height of $10\,\text{m}$ above the ground.

Given that the coefficient of restitution between the ball and the ground is 0.6

(b) show that, after the first bounce at O, the string does not become taut.   [L]

25 A cricket match is played on level ground. Just before the batsman hits the ball, it is $1\,\text{m}$ above the ground and travelling with speed $20\,\text{m}\,\text{s}^{-1}$ in a direction inclined at $\dfrac{\pi}{6}$ below the horizontal. Immediately after the ball has been hit it has speed $25\,\text{m}\,\text{s}^{-1}$ and the direction of motion is reversed. The ball has mass $0.15\,\text{kg}$.

(a) Find the magnitude of the impulse on the ball due to the impact.

Given that the ball first reaches the horizontal ground at a point A, find

(b) the horizontal distance of A from the batsman,

(c) the speed of the ball when it reaches A.

(d) Write down two assumptions which you have made about the forces acting on the ball during its motion. [L]

26 Two particles, *A* and *B*, of masses $2m$ and $3m$ respectively, are moving in a straight line in the same direction on a smooth horizontal plane. The particles collide and, *after* the collision, *A* and *B* continue to move in the same straight line and in the same direction with speeds u and $3u/2$ respectively. Given that the coefficient of restitution between *A* and *B* is $1/5$, find, in terms of *u*, the speed of *A* and the speed of *B* *before* their collision. Find also, in terms of *m* and *u*, the magnitude of the impulse of the force exerted by *B* on *A* during the collision.[L]

27 Three identical smooth spheres, *A*, *B* and *C*, each of mass *m*, lie at rest in a straight line on a smooth horizontal table. Sphere *A* is projected with speed *u* to strike sphere *B* directly. Sphere *B* then strikes sphere *C* directly. The coefficient of restitution between any two spheres is *e*. Find the speeds, in terms of *u* and *e*, of the spheres after these two collisions. Show also that the total loss of kinetic energy is

$$\frac{mu^2}{16}[5 + 2e - 4e^2 - 2e^3 - e^4].$$ [L]

Describe a physical situation that could be modelled by the above problem.

28

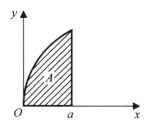

The diagram shows a finite region *A* which is bounded by the curve with equation $y^2 = 4ax$, the line $x = a$ and the *x*-axis. A uniform solid S_1 is formed by rotating *A* through 2π radians about the *x*-axis.

(a) Show that the volume of S_1 is $2\pi a^3$.

(b) Show that the centre of mass of S_1 is a distance $\frac{2a}{3}$ from the origin *O*. [L]

29 Show, by integration, that the centre of mass of a uniform solid hemisphere H, of base radius a, is at a distance $3a/8$ from the centre C of the plane circular base.

The plane face of a second hemisphere K, of radius $a/2$, and made of the same material as H, is stuck to the plane face of H, so that the centres of the two plane faces coincide at C, to form a uniform composite body S. Given that the mass of K is M, show that the mass of S is $9M$, and find, in terms of a, the distance of the centre of mass of the body S from C.

A particle P, of mass M, is attached to a point on the edge of the circular face of H of the body S. The body S with P attached is placed with a point of the curved surface of the part H in contact with a horizontal plane and rests in equilibrium. Find the tangent of the acute angle made by the line PC with the horizontal.   [L]

30 Prove, by integration, that the position of the centre of mass of a uniform solid right circular cone is one quarter of the way up the axis from the base.

From a uniform solid right circular cone of height H is removed a cone with the same base and of height h, the two axes coinciding. Show that the centre of mass of the remaining solid S is a distance

$$\tfrac{1}{4}(3H - h)$$

from the vertex of the original cone.

The solid S is suspended by two vertical strings, one attached to the vertex and the other attached to a point on the bounding circular base. Given that S is in equilibrium, with its axis of symmetry horizontal, find, in terms of H and h, the ratio of the magnitude of the tension in the string attached to the vertex to that in the other string.   [L]

31 Using integration, show that the centre of mass of a uniform thin hemispherical bowl of radius a is at a distance $\tfrac{1}{2}a$ from the centre C of the circular rim of the bowl.

A lid is attached to the bowl to form a closed composite body B. The lid is a thin circular disc of radius a and centre C. The lid is made of the same uniform material as the bowl. Show that the centre of mass of B is at a distance $\frac{1}{3}a$ from C.

The body B has mass M. A particle P, also of mass M, is attached at a point on the circumference of the plane circular face of B. The body is placed with a point of its curved surface in contact with a horizontal plane and rests in equilibrium. Find, to the nearest half degree, the angle made by the line PC with the horizontal. [L]

32 A uniform semi-circular lamina has radius $2a$ and the mid-point of the bounding diameter AB is O.

(a) Using integration, show that the centre of mass of the lamina is at a distance $\dfrac{8a}{3\pi}$ from O.

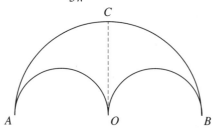

The two semi-circular laminas, each of radius a and with AO and OB as diameters, are cut away from the original lamina to leave the lamina $AOBC$ shown in the diagram, where OC is perpendicular to AB.

(b) Show that the centre of mass of the lamina $AOBC$ is at a distance $\dfrac{4a}{\pi}$ from O.

The lamina $AOBC$ is of mass M and a particle of mass M is attached to the lamina at B to form a composite body.

(c) State the distance of the centre of mass of the body from OC and from OB.

The body is smoothly hinged at A to a fixed point and rests in equilibrium in a vertical plane.

(d) Calculate, to the nearest degree, the acute angle between AB and the horizontal. [L]

33 A uniform triangular lamina XYZ has $XY = XZ$ and the perpendicular distance of X from YZ is h. Prove, by integration, that the centre of mass of the lamina is at a distance $2h/3$ from X.

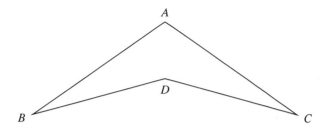

A uniform triangular lamina ABC has $AB = AC = 5a$, $BC = 8a$ and D is the centre of mass of the lamina. The triangle BCD is removed from the lamina, leaving the plate $ABDC$ shown in the diagram.

(a) Show that the distance of the centre of mass of the plate from A is $5a/3$.

The plate, which is of mass M, has a particle of mass M attached at B. The loaded plate is suspended from C and hangs in equilibrium.

(b) Prove that in this position CB makes an angle of $\arctan(1/9)$ with the vertical. [L]

34 A length of uniform thin wire is bent to form a semicircular arc of radius r. Use integration to show that the centre of mass is at a distance $\dfrac{2r}{\pi}$ from the diameter joining the end points.

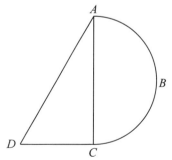

The diagram shows a plane frame made from uniform thin wire. The frame consists of straight wires AC, CD and DA and a semicircular wire ABC. The semicircular arc ABC is of radius

8 cm, angle $ACD = 90°$, and CD is of length $2x$ cm. Given that the centre of mass of the frame lies on AC, show that

$$3x^2 = 64$$

[L]

35 Use integration to show that the centre of mass of a uniform solid right circular cone, of vertical height h and base radius r, is at a distance $\frac{3}{4}h$ from the vertex V of the cone.

A uniform composite body consists of this cone and a solid hemisphere, of radius r, whose plane faces coincide. The centre of mass of the composite body is at G. The diagram shows a plane section of the body taken through V and G.

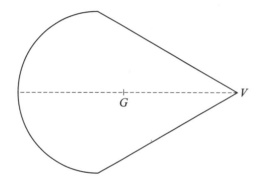

(a) Show that $VG = \dfrac{3h^2 + 8hr + 3r^2}{4(h + 2r)}$.

(b) If $VG = h$, find the semi-vertical angle of the cone.

(c) If $VG = r$, deduce that $3h = r(\sqrt{19} - 2)$. [L]

36 A uniform triangular lamina ABC has $\angle ABC = 90°$ and $AB = c$. Using integration, show that the centre of mass of the lamina is at a distance $\frac{1}{3}c$ from BC.

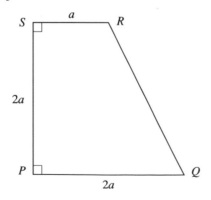

The diagram shows a uniform lamina in which

$$PQ = PS = 2a, SR = a, \angle SPQ = \angle RSP = 90°.$$

The centre of mass of this lamina is G.

(a) Show that the distance of G from PS is $\frac{7}{9}a$.

(b) Find the distance of G from PQ.

The point R of the lamina is smoothly hinged at a fixed point so that the lamina can rotate freely in a vertical plane. The mass of the lamina is $3M$.

(c) Find, in terms of M, g and a, the magnitude of the couple required to keep the lamina in equilibrium with the edge RQ vertical. [L]

37 Use integration to show that the centre of mass of a uniform semi-circular lamina, of radius a, is at a distance $4a/3\pi$ from O the mid-point of its straight edge.

A semi-circular lamina, of radius b and with O as the mid-point of its straight edge, is removed from the first lamina. Show that the centre of mass of the resulting lamina is at a distance $\bar{x}$ from O, where

$$\bar{x} = \frac{4}{3\pi}\frac{(a^2+ab+b^2)}{(a+b)}.$$

Hence find the position of the centre of mass of a uniform semi-circular arc of radius a. [L]

38

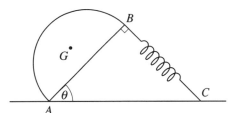

The diagram shows a uniform solid hemisphere of mass m, base radius a and with centre of mass G. The hemisphere is freely pivoted at a point A on the circumference of its plane face. The line AB is a diameter of the plane face and a light spring connects B to a point C on the horizontal plane through A. The system is in equilibrium with A, B, C and G in the same vertical plane, BC is perpendicular to AB and AB is inclined to AC at an angle θ, where $\tan\theta = 4/3$. Show that the thrust in the spring has magnitude $3mg/20$.

Given that the modulus of the spring is $3mg/4$, find the natural length of the spring.   [L]

39

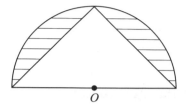

The diagram shows a cross-section of a solid formed by the removal of a right circular cone, of base radius a and height a, from a uniform solid hemisphere of base radius a. The plane bases of the cone and the hemisphere are coincident, both having centre O. Show that G, the centre of mass of the solid, is at a distance $a/2$ from O.

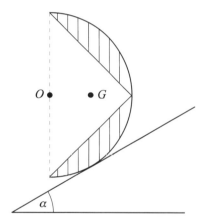

The second diagram shows a cross-section of the solid resting in equilibrium with a point of its curved surface in contact with a rough inclined plane of inclination α. Given that O and G are in the same vertical plane through a line of greatest slope of the inclined plane, and that OG is horizontal, show that $\alpha = \pi/6$. Given that $\alpha = \pi/6$, find the smallest possible value of the coefficient of friction between the solid and the plane.   [L]

40 A smooth horizontal rail is fixed at a height $3h$ above rough horizontal ground. A uniform rod AB, of mass M and length $6h$, is placed in a vertical plane perpendicular to the rail with the end A resting on the ground. The distance $AC = 5h$, where C is the point of contact between the rail and the rod. Show that the force exerted by the rail on the rod is of magnitude $12Mg/25$.

Given that equilibrium is limiting, find the coefficient of friction between the rod and the ground and show that the force exerted by the ground on the rod is of magnitude $17Mg/25$.

Find, in terms of M and g, the greatest magnitude of the horizontal force which could be applied to the rod at A without disturbing equilibrium.   [L]

41 A uniform straight rod AB, of mass M and length $2l$, rests in limiting equilibrium with the end A on rough horizontal ground and the end B against a smooth vertical wall. The vertical plane containing AB is at right angles to the wall and the coefficient of friction between the rod and the ground is $\frac{1}{3}$.

(a) Show that AB is inclined at $\arctan\left(\frac{3}{2}\right)$ to the horizontal.

With AB in the same position, a horizontal force of magnitude kMg is applied to the mid-point of the rod towards the wall so that the line of action of this force is at right angles to the wall. Given that equilibrium is limiting with the end A on the point of moving towards the wall, calculate

(b) the value of k,

(c) the magnitude of the resultant force exerted by the rod on the ground, giving your answer in terms of M and g.   [L]

42 A uniform rod AB, of length $2a$ and mass m, rests in equilibrium with its lower end A on a rough horizontal floor. Equilibrium is maintained by a horizontal elastic string, of natural length a and modulus λ. One end of the string is attached to B and the other end to a point vertically above A. Given that θ, where $\theta < \pi/3$, is the inclination of the rod to the horizontal, show that the magnitude of the tension in the string is $\frac{1}{2}mg\cot\theta$.

Prove also that

$$2\lambda = (mg\cot\theta)/(2\cos\theta - 1).$$

Given that the system is in limiting equilibrium and that the coefficient of friction between the floor and the rod is $2/3$, find $\tan\theta$. Hence show that $\lambda = 10\,mg/9$.   [L]

43

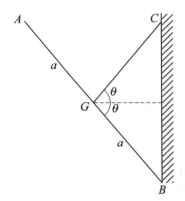

The diagram shows a uniform rod AB of mass M and length $2a$ with the end B in contact with a rough vertical wall. The rod is held in equilibrium by a light inextensible string of length a with one end attached to the mid-point G of the rod and the other end attached to a point C of the wall vertically above B. The string and the rod each make an acute angle θ with the horizontal.

(a) Show that μ, the coefficient of friction between the wall and the rod, satisfies the relation $\mu \geqslant \tan \theta$.

When a force of magnitude $\frac{1}{2} Mg$ acts at A, in the direction from A to C, the system remains in equilibrium.

(b) Find, in terms of M, g and θ, expressions for the horizontal and vertical components of the force acting on the rod at B.

(c) Given that $\mu = 0.75$, deduce that $\tan \theta \leqslant \frac{1}{2}$. [L]

44 A smooth cylinder of radius a is fixed on a rough horizontal table with its axis parallel to the table. A uniform rod ACB of length $6a$ and mass M rests in limiting equilibrium with the end A on the table and the point C touching the cylinder. The vertical plane containing the rod is perpendicular to the axis of the cylinder and the rod makes an angle 2θ with the table.

(a) Show that the magnitude of the force exerted by the cylinder on the rod is

$$3Mg \cos 2\theta \tan \theta.$$

(b) Show also that μ, the coefficient of friction between the rod and the table, is given by

$$\mu(\cot \theta - 3 \cos^2 2\theta) = 3 \sin 2\theta \cos 2\theta. \qquad \text{[L]}$$

45

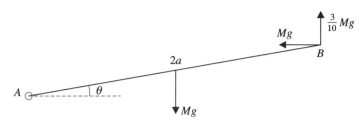

A uniform straight rod AB has mass M and length $2a$. The end A is smoothly hinged at a fixed point so that the rod can turn freely in a vertical plane. Horizontal and vertical forces of magnitudes Mg and $\frac{3}{10}Mg$ respectively are applied to the end B. These forces and the weight of the rod are shown in the diagram above. The rod rests in equilibrium at an angle θ to the horizontal.

(a) By taking moments about A, find the value of $\tan\theta$.

(b) Calculate the magnitude, in terms of M and g, and the direction, to the nearest degree, of the force exerted by the hinge on the rod AB.   [L]

46 A uniform ladder of length $4l$ and mass M rests with one end A on rough horizontal ground and the other end B against a smooth vertical wall. The vertical plane containing AB is at right angles to the wall. The coefficient of friction between the ladder and the ground is $\frac{1}{5}$. A particle of mass $2M$ is attached to the ladder at C where $AC = 3l$.

Given that equilibrium is limiting, show that the ladder is inclined at an angle $\arctan\left(\frac{10}{3}\right)$ to the horizontal.

The ladder is moved to a similar position where the wall is rough. The ladder rests at an angle $\arctan\left(\frac{10}{3}\right)$ to the horizontal and the coefficients of friction between the ladder and the wall and the ladder and the ground are both $\mu(\neq\frac{1}{5})$. The particle of mass $2M$ is moved to the top of the ladder. Given that equilibrium is limiting, show that

$$\mu^2 + 20\mu - 5 = 0.$$ [L]

Obtain the value of μ to 3 decimal places.

Describe a physical situation that this problem could be used to model.

State the assumptions made in producing this model.

Interpret your final answer in terms of the original situation.

47

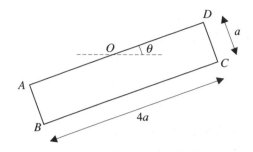

The diagram shows a uniform rectangular lamina $ABCD$, free to rotate in a vertical plane about a smooth horizontal axis through O, the mid-point of AD. The lamina is of mass $2m$ and the sides AB and BC are of length a and $4a$ respectively. The angle θ is the inclination of OD to the horizontal. Particles of mass $5m$, $2m$ and $4m$ are attached at the points A, B and D respectively.

Find, in terms of m, a, g and θ, an expression for the anti-clockwise moment about O of the four weights.

Given that the system is in equilibrium, show that $\tan \theta = 2$. When a force of magnitude P is applied at the point C and in the direction $\overrightarrow{CB}$ the system rests in equilibrium with AD horizontal and uppermost. Show that $P = 6mg$ and, in this position, find the magnitude of the reaction at O.   [L]

48 A uniform rod AB, of mass m and length l, is smoothly jointed at the end A to a fixed straight horizontal wire AC. The end B is attached by means of a light inextensible string, also of length l, to a small ring, of mass m, which can slide on the wire, the coefficient of friction between the ring and the wire being μ. The rod is in equilibrium in the vertical plane through the wire. Given that α is the inclination of the string to the horizontal, show that the tension in the string is of magnitude $mg/(4 \sin \alpha)$. Show also that

$$\tan \alpha \geqslant 1/(5\mu). \qquad \text{[L]}$$

49 A uniform ladder, of mass M and length l, rests in a vertical plane with one end against a smooth vertical wall, the wall being perpendicular to the vertical plane through the ladder. The other end of the ladder rests on horizontal ground, the coefficient of friction between the ladder and the ground being

$1/4$. The ladder is inclined at an angle θ to the horizontal, where $\tan \theta = 24/7$. A man of mass $10M$ climbs up the ladder. Show that the man can reach a height of $6l/7$ above the ground before the ladder begins to slip. [L]

What assumptions have you made in modelling the above physical situation in order to obtain the mathematical model you solved?

How could this model be refined so that it becomes a better model of the physical situation?

50 A uniform ladder, of mass 45 kg and length 4 m, rests in a vertical plane with one end against a smooth vertical wall, the wall being perpendicular to the vertical plane through the ladder. The other end of the ladder rests on horizontal ground, the coefficient of friction between the ladder and the ground being 0.4. The ladder is inclined at an angle θ to the horizontal where $\tan \theta = 1.5$. A man of mass 60 kg climbs up the ladder When he reaches a point x m from the bottom of the ladder, measured along the ladder, the ladder slips.

Model the above situation and hence obtain the value of x. State clearly the assumptions you have made in producing your model. How can your model be refined so as to take into account further properties of the actual situation?

51 A uniform metal rod AB of length 3 m and mass 20 kg is smoothly hinged at A to a vertical wall. It is held in a horizontal position by a cord fixed at the end B of the rod. The other end of the cord is fixed to a point of the wall vertically above A so that the cord makes an angle of $30°$ to the horizontal. A lamp of mass 30 kg is suspended from B.

(a) Model this situation and hence obtain an estimate for the tension in the cord. State your modelling assumptions.

The breaking tension of the cord is 1960 N.

(b) It is required to hang an additional piece of equipment of mass 90 kg from the rod. Find the greatest possible distance from A that this can be hung without the cord breaking.

Examination style paper

M2

Whenever a numerical value of g is required, take $g = 9.8\,\text{m}\,\text{s}^{-2}$

1 The amplitude of oscillation of a particle describing simple harmonic motion is 3 m. When the particle is 1 m from the centre of the oscillation its speed is $4\,\text{m}\,\text{s}^{-1}$. Find the periodic time of the motion.

(4 marks)

2 An ice puck of mass 0.2 kg is moving on a horizontal ice rink with velocity $(10\mathbf{i}-4\mathbf{j})\,\text{m}\,\text{s}^{-1}$ when it is struck by a hockey stick. The velocity of the puck after the impact is $(16\mathbf{i}+4\mathbf{j})\,\text{m}\,\text{s}^{-1}$. Model this situation and obtain an estimate for the magnitude of the impulse applied to the puck by the stick. State clearly the assumptions you made in obtaining your model.

(7 marks)

3 A simple pendulum consists of a small bob P of mass m attached by a light inextensible string to a fixed point O. The pendulum swings through an angle $\alpha\,(< \frac{\pi}{2})$ either side of the vertical through O. Show that when OP makes an angle θ with the vertical the tension in the string is

$$mg(3\cos\theta - 2\cos\alpha)$$

(8 marks)

4 A particle P is moving along the positive x-axis. At time t seconds the particle is at a distance x metres from the origin O, its speed is $v\,\text{m}\,\text{s}^{-1}$ and its acceleration is $(x + 2)\,\text{m}\,\text{s}^{-1}$. Given that when $t = 0$, the velocity of P is $2\,\text{m}\,\text{s}^{-1}$ in the direction Ox, and $x = 0$,
(a) show that $v = x + 2$,
(b) find x as a function of t.

(11 marks)

5 It is required to estimate the force exerted by a hammer thrower. The athlete is regarded as the vertical axis of a conical pendulum. The hammer is treated as a particle of mass 6 kg attached to a light inextensible wire of length 1 m. The length of the wire together with the

athlete's arm is estimated to be 2 m. The particle has angular speed $5 \, \text{rad s}^{-1}$.

(a) Calculate the magnitude of the force exerted by the hammer thrower.

(b) Find the angle between the wire and the vertical.

(c) Discuss briefly the assumptions made in this model.

(12 marks)

6 Using Hooke's Law, show by integration that the work done in stretching a light elastic string of natural length l and modulus of elasticity λ, from length l to $(l + e)$ is $\dfrac{\lambda e^2}{2l}$.

A light elastic string of natural length 4 metres and modulus of elasticity $4mg$ has one end attached to a fixed point A. A particle P, of mass m, is attached to the other end of the string.

(a) Show that, when P hangs in equilibrium, the length of AP is 5 m.

The particle is released from rest at A and falls vertically.

(b) Find the greatest length of the string.

(13 marks)

7 Three small smooth spheres A, B and C have equal radii and masses m, $2m$ and $4m$ respectively. They are free to move along a smooth straight horizontal groove with B between A and C. The coefficient of restitution between any two spheres is e $(e < \frac{1}{2})$. Spheres B and C are at rest and A is projected towards B with speed u.

(a) Show that after the first impact the speeds of A and B are $\frac{1}{3}(1 - 2e)u$ and $\frac{1}{3}(1 + e)u$ respectively.

(b) Find the speeds of B and C after the second impact.

(c) Hence show that there is a third impact.

(14 marks)

8 *(a)* Show, by integration, that the centre of mass of a uniform solid right circular cone of height h is at a distance $\frac{3}{4}h$ from the vertex.

Two uniform solid right circular cones, each of base radius a, have the same density and are of heights h and λh respectively where $\lambda > 1$. These cones are joined together, so that their circular bases coincide, to form a spindle as shown in the diagram.

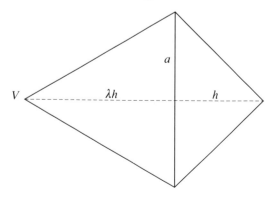

(b) Show that the centre of mass of this spindle is at a distance $\frac{1}{4}h(3\lambda + 1)$ from the vertex V of the larger cone.

Given that $a = h$:

(c) show that the spindle can rest in equilibrium with the curved surface of the smaller cone in contact with the horizontal plane provided that $\lambda \leqslant 5$.

(15 marks)

9 A uniform ladder AB, of mass 30 kg, rests in equilibrium with the end A in contact with rough horizontal ground and the end B against a smooth vertical wall. The vertical plane containing AB is at right angles to the wall and AB is inclined at $60°$ to the horizontal. The coefficient of friction between the ladder and the ground is μ.

(a) Find the magnitude of the force exerted by the wall on the ladder.

(b) Show that $\mu \geqslant \frac{1}{6}\sqrt{3}$.

A load of mass m is attached to the ladder at B.

(c) Given that $\mu = \frac{1}{5}\sqrt{3}$ and that the equilibrium is limiting, find m.

(16 marks)

Answers

The University of London Examination and Assessment Council accepts no responsibility whatsoever for the accuracy or method of working in the answers given for examination questions.

Exercise 1A

1 $21 \, \text{m s}^{-1}$ 2 $48 \, \text{m s}^{-1}$ 3 $9 \, \text{m s}^{-1}$, $9 \, \text{m}$

4 $29 \, \text{m s}^{-1}$, $24 \, \text{m}$ 5 $7\frac{1}{2} \, \text{s}$, $29\frac{1}{4} \, \text{m}$

6 $19 \, \text{m s}^{-1}$

7 (a) $t = 1, 3$ (b) $4 \, \text{m}$

8 $1 \, \text{m s}^{-1}$, $\dfrac{\pi}{4} \, \text{m}$

9 $t^3 + 2t + 2$, 35 10 $x = e^{-t} + 5t + 4$

11 $t = 5$

12 $(-3 \ln 7 + 18) \, \text{m}$

Exercise 1B

1 $v^2 = u^2 + 2fx$ 2 $8 \, \text{m s}^{-1}$

3 $v^2 = 1600 - 40x - 2x^2$, $x = 20$

4 $\sqrt{\left(\dfrac{a^3 2\lambda}{3}\right)}$ 5 $x = \exp\left(\dfrac{v^2}{2k}\right)$ 6 2

7 (a) $v = x + 3$ (b) $x = 3e^t - 3$

8 (a) $v = \dfrac{2}{x}$ (b) $x = \sqrt{4t + 1}$

9 (a) $v = -e^x$ (b) $x = \ln\left(\dfrac{u}{ut + 1}\right)$

10 $t = \frac{1}{4}\sqrt{15}$ 11 $2\sqrt{13} \, \text{m s}^{-1}$

12 $x = -\ln(1 - t)$

14 $1.287 \, \text{m s}^{-1}$

Exercise 2A

1 (a) $3.68 \, \text{m s}^{-1}$ (b) $15.9 \, \text{m s}^{-1}$
   (c) $t = 2\pi \, \text{s}$ (d) $22.9 \, \text{m}$ (e) $16\pi \, \text{s}$

2 (a) $3.736 \, \text{m s}^{-1}$ (b) $5.099 \, \text{m s}^{-1}$

3 (a) $54 \, \text{m s}^{-1}$ (b) $6.9 \, \text{m}$

4 (a) $4 \, \text{m s}^{-1}$ (b) $28 \, \text{m}$
   (c) $\frac{5}{3} \, \text{s}$ (d) $25.9 \, \text{m}$

5 (a) $9.90 \, \text{m s}^{-1}$ (b) $1.70 \, \text{s}$

6 (a) $28 \, \text{N s}$ (b) $0.549 \, \text{N s}$
   (c) $15.5 \, \text{N s}$ (d) $0.6 \, \text{N s}$

7 (a) $66.7 \, \text{J}$ (b) $8.13 \, \text{J}$
   (c) $7.39 \, \text{J}$

8 (a) $32 \, \text{N s}$ (b) $32 \, \text{m s}^{-1}$ (c) $6 \, \text{s}$

9 $\dfrac{0.8}{3}(x^2 + x)$, $14.4 \, \text{N m}$

Exercise 2B

1 (a) $1.67 \times 10^{-8} \, \text{N}$ (b) $1.67 \times 10^{-12} \, \text{N}$
   (c) $1.67 \times 10^{-18} \, \text{N}$

2 $0.534 \, \text{kg}$

3 $9.78 \, \text{m s}^{-2}$ Earth can be modelled as a particle

4 $5.96 \times 10^{24} \, \text{kg}$ 5 $1.46 \times 10^2 \, \text{kg}$

6 $1.62 \, \text{m s}^{-2}$ Particle model

Exercise 2C

1 $7.5 \, \text{N}$ 2 $0.0625 \, \text{m}$ 3 $0.563 \, \text{m}$

4 $8 \, \text{N}$ 5 $180 \, \text{N}$ 6 $3.123 \, \text{m}$

7 $12.8 \, \text{cm}$ 8 $0.94 \, \text{m}$, $1.27 \, \text{m}$

9 $58.8 \, \text{N}$ 10 $1.96 \, \text{m}$

11 11.2 N, 8.4 N

12 Ball – dimensions small compared with length of cord, particle model. Cord – assume light, light elastic string model. 3.68 N, 61.3 N

13 35.2°, 13.9 N

14 (a) 15.1 N (b) 0.603 m

15 (a) Particle, elastic spring, 44.6 N
    (b) 0.172

16 (a) $12\,\text{m s}^{-2}$ (b) $12\,\text{m s}^{-2}$

18 $10\,\text{m s}^{-2}$ upwards

Exercise 2D

1 $\dfrac{3\pi}{10}$, $1.89\,\text{m s}^{-1}$ 2 $2.5\,\text{m s}^{-1}$

3 $\frac{5}{6}$ m, $1.81\,\text{m s}^{-1}$ 4 2π s, 10 m

5 $11.9\,\text{m s}^{-1}$ 6 $0.5\,\text{m s}^{-1}$, $0.433\,\text{m s}^{-1}$

7 $\frac{5}{2}$m, $x = \frac{5}{2}\sin 2t$

8 $\dfrac{3}{4\pi}$m, 0.0475 m   9 2.53 m, 1.22 s

10 3.29 J Piston moves in a straight line and can be modelled as a particle.

11 Particle model, $0.785\,\text{m s}^{-1}$, 0.912 s

12 5 m, $\dfrac{2\pi}{3}$ s (a) $15\,\text{m s}^{-1}$ (b) 0.280 s
    (c) 0.175 s

13 2 m, $\dfrac{2\pi}{5}$ s, 0.959 m, 0.528 s

14 0.598 s   15 1.462 s

16 2.07 pm   10.33 pm

Exercise 2E

1 (a) 1.42 s (b) 2.29 s (c) 1.68 s

2 (a) 0.345 m (b) 0.484 m (c) 0.248 m

3 0.993 m   4 $\dfrac{T}{\sqrt{2}}$ 5 $9L$

6 Cord is inelastic; metal bob is small and heavy compared with cord; amplitude of oscillations is small. 2.84 s

7 Rope is long compared with height of boy; rope is light compared with weight of boy; rope is inelastic. 3.04 m

Exercise 2F

1 0.7255 s, 0.289 m

2 $x = 0.129 \sin 12.9t$, $21.5\,\text{m s}^{-1}$

3 0.444 s, $2.828\,\text{m s}^{-1}$

4 (a) Particle, light elastic string, smooth horizontal surface
    (b) $3.79\,\text{m s}^{-1}$ (c) 1.55 s
    (d) Smooth cylinder $\Rightarrow$ frictional force negligible. The answer to (b) would be smaller.

5 0.5 m, $2.11\,\text{m s}^{-1}$, 0.25 m

6 (a) Spring light compared with piston; model spring as light elastic spring; assume piston can be modelled as a particle.
    (b) 0.251 s, $2.5\,\text{m s}^{-1}$
    (c) Cylinder unlikely to be perfectly smooth – include frictional force.

7 $1.22\,\text{m s}^{-1}$, 0.976 s

8 0.5 m, 1.72 s, $1.83\,\text{m s}^{-1}$, 1.25 J

Exercise 2G

1 0.49 m, 1.40 s, $2.24\,\text{m s}^{-1}$

2 0.513 s, 0.163 m

3 45 N, 0.8 m

4 (a) 1.25 m (b) 0.25 m (c) $1.5\,\text{m s}^{-1}$

5 Man – particle; spring – light elastic; assume mass of man large compared with spring; assume motion is in a straight line. 0.02 m, 0.284 s

6 Piston – particle; spring – light elastic; assume mass of piston large compared with that of spring, dimensions of piston

small compared with length of spring.
0.196 s, 0.9287 m s^{-1}
Include frictional force

7 0.295 s, 0.503 s   **8** 0.0142 m, 0.468 s

9 (a) Climber – particle; rope – light elastic
string; assume rope long and light
compared with height and weight of
climber, initial speed of climber is
0 m s^{-1}

     (b) 14 m s^{-1} (c) 13.0 m   (d) 1.75 s

Exercise 3A

1 28.6 rev min^{-1} **2** 0.419 rad s^{-1}

3 (a) 10 rev min^{-1} (b) 1.05 rad s^{-1}

4 (a) 7.5 rev min^{-1} (b) 0.785 rad s^{-1}

5 12.5 m s^{-1} **6** 0.0785 m s^{-1}

7 41.9 m s^{-1} **8** 30 rad s^{-1}, 1.26 s

9 1.8 rad s^{-1} **10** 0.4 rad s^{-1}

11 (a) 1.6 rad s^{-1} (b) 1.2 m s^{-1}

Exercise 3B

1 0.206 m s^{-2} **2** 200 m s^{-2}

3 (a) 10 rad s^{-1} (b) 15 m s^{-1}

4 5.6 m s^{-2} **5** 1.48 m s^{-2}

6 0.0337 m s^{-2}

Exercise 3C

1 83.3 N   **2** 2.5 m s^{-1}

3 3.30 rad s^{-1} **4** 0.638

5 7.2 N   **6** 2.5 m

7 (a) 10 N   (b) 17.3 rad s^{-1}

9 13 : 10   **11** 3.91 rad s^{-1}

12 46.1 N towards centre

13 Particle model, $2mg$, $\frac{1}{2}\sqrt{(6gr)}$

14 Particle, smooth surface, 3.13 m s^{-1}

15 8.32 N, 4.16 rad s^{-1}

16 48 N, 0.476 m

17 (a) 21.6 N   (b) 8.8 N

18 (a) $\dfrac{lg \sin^2 \phi}{4 \cos \phi}$ **19** $2\sqrt{\left(\dfrac{g}{3l}\right)}$

20 Particle model, $\dfrac{5mg}{3}$, $2\sqrt{(gl)}$

     (b) The tensions in the two parts of the
string are the same.

21 $\dfrac{7mg}{12}$ **22** 17 : 15   **23** $2\sqrt{\left(\dfrac{g}{3l}\right)}$

Review exercise 1

1 53.2 m

2 $a = 6 + 2t$, $x = 3t^2 + \frac{1}{3}t^3 - \frac{44}{3}$, 102 m

3 20 m

4 (a) $1\frac{5}{27}$ m   (b) 2 s   (c) 4 m s^{-1}

5 (b) $43\frac{1}{3}$ m

7 (a) $v = 1 + 2\cos 2t$ (b) $f = -4\sin 2t$

8 (b) 115 m   (c) 130 J   (d) 8 N

9 (b) $v = 9.8(e^t - 1)$

     (c) Assumed small enough to be
modelled as a particle.

10 $\sqrt{\left(\dfrac{3gh}{2}\right)}$

11 (a) 3.5 s   (b) $\dfrac{19.6}{k}(1 - e^{-\frac{kt}{2}})$

12 (a) 21 N   (b) 47 m s^{-1}

13 (a) 12 J   (b) 7.75 m s^{-1}

14 655 N, Earth – particle, same mass,
situated at centre of earth; object –
particle. 7570 m s^{-1}

15 3.58×10^4 km   **16** 1.93 h

17 6.52 m s^{-1}, 2970 km

18 $l + \dfrac{mgl}{\lambda}$ **19** 250 mm   **20** 0.3 m

21 118 N   (a) 25.5 N   (b) 14.7 N   (c) 1.35 m

22 $\dfrac{79c}{36}$

23 (a) 19.6 N   (b) 43.1 N

24 12 : 5, 11.76 N, 106 N

25 (a) $\dfrac{l}{4}$ **26** $x = a \sin nt$

27 (a) $-an \sin nt$, $-an^2 \cos nt$ (b) $a^2 n^2$
(c) $an^3 \sin nt = -n^2 v$, S.H.M.

28 $\dfrac{\sqrt{3}}{2}\,\mathrm{m\,s^{-1}}$

29 (a) $-k\omega\pi$ (b) $\dfrac{k\pi}{4\omega}$

30 (a) 2.5 m (b) $\dfrac{\pi}{4}$ s (c) $-10\,\mathrm{m\,s^{-1}}$
(d) $20\,\mathrm{m\,s^{-2}}$ (e) $\dfrac{2\pi}{3}$ s

31 $an \cos nt$, $-an^2 \sin nt = -n^2 s$, $a = 0.15$,
$n = 4$, 0.785 s

32 (a) $\pi\sqrt{\left(\dfrac{2l}{g}\right)}$ $\dfrac{l}{4}$

33 (a) 0.945 h
(b) Particle moving in a straight line
with S.H.M.

34 3.01 s **35** $2\pi\sqrt{\left(\dfrac{l}{g}\right)}$

36 3, $\dfrac{3l}{4}$ **37** $2a\sqrt{\left(\dfrac{g}{l}\right)}$

38 $\dfrac{l}{10}$, $2\pi\sqrt{\left(\dfrac{l}{5g}\right)}$

39 51.2 N, $17.3\,\mathrm{rad\,s^{-1}}$

41 (a) $mg \sec \beta$ (b) $\dfrac{g}{a \cos \beta}$

42 $\frac{3}{4}$ m **44** $\dfrac{4l}{3}$

46 (i) (a) $m^2 g^2 + \dfrac{m^2 v^4}{d^2}$ (b) $\dfrac{gd}{v^2}$
(ii) Particle model, air resistance.

47 $19.2\,\mathrm{m\,s^{-1}}$

48 (a) $\dfrac{3l}{2}$ (c) $2mg$ (d) $2\pi\sqrt{\left(\dfrac{l}{g}\right)}$

50 1.60 h

Exercise 4A
1 (a) $8.40\,\mathrm{m\,s^{-1}}$ (b) 127 N
2 $3.13\,\mathrm{m\,s^{-1}}$ **3** $4.43\,\mathrm{m\,s^{-1}}$, 3.13 N
4 (a) $10.2\,\mathrm{m\,s^{-1}}$ (b) $5.12\,\mathrm{m\,s^{-1}}$ No
5 (b) $2\sqrt{(gl)}$
6 As a particle (b) $ag(2 + \sqrt{3})$
(c) $0.232\,mg$ towards O
7 $5.64\,\mathrm{m\,s^{-1}}$

Exercise 4B
1 $4.43\,\mathrm{m\,s^{-1}}$, 3.68 N
2 (a) $9.9\,\mathrm{m\,s^{-1}}$ (b) 206 N, 147 N
3 (a) $7.78\,\mathrm{m\,s^{-1}}$ (b) Moves as a projectile
until string is taut again.
4 (a) $6.64\,\mathrm{m\,s^{-1}}$ (b) $64.5°$
5 (a) As a particle (b) $4.18\,\mathrm{m\,s^{-1}}$, 168 N
6 $3mg(1 - \cos \theta)$, $110°$
7 $\sqrt{(3gl)}$
8 (a) $\sqrt{(gr)}$, $\frac{1}{2}mg$ (b) $\sqrt{(2gr)}$
9 $\sqrt{(ag)}$
11 (a) Particle (b) $\sqrt{(ag)}$ (c) $\sqrt{(5ag)}$
(d) $2a$

Exercise 4C
1 1 J **2** 1.5 J
3 (a) 0.72 J (b) 4.62 J
4 (a) 1.25 J
(b) 2.5 J They are different although
change in length is the same.
5 2.49 J **6** $\frac{1}{2}mgl$ **7** $\frac{3}{2}mgl$
8 $\frac{1}{4}mgl$

Exercise 4D
1 $1.37\,\mathrm{m\,s^{-1}}$
2 (a) $1.73\,\mathrm{m\,s^{-1}}$ (b) $1.63\,\mathrm{m\,s^{-1}}$
3 (a) 3.34 m (b) $3.82\,\mathrm{m\,s^{-1}}$
4 0.180 m
7 (b) $4l$

8 $0.665\,\text{m}$, $1.25\,\text{m}\,\text{s}^{-1}$

9 $48.1\,\text{N}$, $2.90\,\text{m}\,\text{s}^{-1}$

10 Assume stone is small, model as particle; assume elastic light.

   (b) $20.1\,\text{m}\,\text{s}^{-1}$

   (c) Answer different: gravitational P.E. would have to be included.

Exercise 5A

1 $(12\mathbf{i} - 2\mathbf{j})\,\text{m}\,\text{s}^{-1}$, ice hockey

2 $(-\mathbf{i} - 2\mathbf{j})\,\text{N}\,\text{s}$ **3** $(\mathbf{i} + 9\mathbf{j})\,\text{m}\,\text{s}^{-1}$

4 $I = 18$ **5** $I = 36$

6 $24\,\text{N}\,\text{s}$, a lump of putty   **7** $52.7\,\text{N}\,\text{s}$

8 $\frac{3}{7}\,\text{m}\,\text{s}^{-1}$, original direction of A. An engine and a carriage of a train set which couple on colliding.

9 $3\frac{1}{2}\mathbf{i}\,\text{m}\,\text{s}^{-1}$

10 $5\,\text{m}\,\text{s}^{-1}$, direction reversed

Exercise 5B

1 (a) $e = \frac{1}{2}$ (b) $e = \frac{1}{4}$ (c) $e = \frac{2}{3}$

2 (a) $v_1 = 0$, $v_2 = 4$ (b) $v_1 = 3$, $v_2 = 4$

   (c) $v_1 = 2\frac{1}{2}$, $v_2 = 5$

   (d) $v_1 = -2$, $v_2 = 1$ (e) $v_1 = -1$, $v_2 = 0$

3 (a) $4\,\text{m}\,\text{s}^{-1}$ (b) $e = \frac{1}{4}$

   (c) Spheres of same size and can be modelled as particles.

4 (a) $\frac{3}{5}$ (b) $7.5\,\text{J}$

5 $e > \frac{1}{4}$

6 Two snooker balls on a snooker table. Assume balls may be treated as particles and that the table is perfectly smooth.

Exercise 5C

1 (a) $e = \frac{1}{2}$ (b) $e = \frac{3}{4}$

2 (a) $v_1 = 4$ (b) $v_2 = 2.5$

3 (a) $u_1 = 10$ (b) $u_2 = 8$

4 (a) $e = \frac{3}{4}$ (b) $12.6\,\text{J}$

   (c) Snooker ball striking a cushion at right angles. Puck on ice rink striking boundary normally.

5 $e = \dfrac{1}{\sqrt{2}}$ **7** $e = \frac{1}{2}$ **8** $h = 128$

9 $e = \frac{3}{5}$. Small rubber ball dropped on to smooth pavement or ice rink.

Exercise 5D

1 (a) $x = 2$, $y = 6$, $v = 1\frac{1}{2}$, $w = 4\frac{1}{2}$

   (b) $x = \frac{1}{2}$, $y = 2$, $v = \frac{3}{4}$, $w = 1\frac{1}{2}$

2 (a) $6\,\text{m}\,\text{s}^{-1}$

4 (a) $1\,\text{m}\,\text{s}^{-1}$, $2\,\text{m}\,\text{s}^{-1}$ and $8\,\text{m}\,\text{s}^{-1}$ respectively.

   (b) No, since (speed C) > (speed B) > (speed A)

5 Three snooker balls on a smooth horizontal table.

6 (b) $\frac{2}{9}$

Exercise 6A

1 $(2, \frac{4}{3})$ **2** $(\frac{3}{4}, \frac{3}{10})$ **3** $(\frac{124}{75}, \frac{254}{105})$

4 $(\frac{3}{5}b, 0)$ **5** $(\frac{1}{3}a, \frac{1}{3}h)$ **6** $(\frac{9}{16}, \frac{7}{10})$

7 $\left(\dfrac{4r}{3\pi}, \dfrac{4r}{3\pi}\right)$ **8** $\left(\dfrac{\pi}{2} - 1, \dfrac{\pi}{8}\right)$

9 $\left(\dfrac{1}{e - 1}, \dfrac{e + 1}{4}\right)$

10 $\dfrac{28}{3\pi}$ **11** $(\frac{1}{2}, \frac{2}{5})$

Exercise 6B

1 $(3, 0)$ **2** $(1, 0)$ **3** $\left(\dfrac{\pi}{2}, 0\right)$ **4** $(0, \frac{1}{2})$

5 $(\frac{5}{6}, 0)$ **6** $(\frac{7}{8}, 0)$ **7** $(\frac{5}{8}, 0)$

8 $(2\ln 2, 0)$ **9** $(1\frac{2}{7}, 0)$

10 $\left(\dfrac{1}{2}\dfrac{(e^2 + 1)}{(e^2 - 1)}, 0\right)$ **11** $\left(\dfrac{e^2 - 3}{2(e^2 - 1)}, 0\right)$

12 $\dfrac{3(h^2 + 4lh + 2l^2)}{4(h + 3l)}$ **13** $\sqrt{6} : 1$

14 $3\frac{3}{14} = 3.21$ cm   **16** 2.33 cm

Exercise 6C
1 $69.4°$ **4** $53.1°$ **5** $30°$
6 $r\sqrt{3}$ **7** $\frac{75}{33}r$ from joint face

Exercise 6D
2 84.9 N   **3** $45°$
4 (a) 1176 N, 467 N (b) 0.505
5 $\frac{2}{3}$ **6** (a) $36.9°$ (b) $\frac{1}{2}$
7 11.4 N
8 (a) $45°$
  (b) A uniform ladder against a fairly smooth wall
  (c) Take into account roughness of wall; model ladder by non-uniform rod.

Review exercise 2
1 (a) 3.33 m s^{-1}, 0.566 m
  (b) Ropes may be modelled by a light inextensible string. Weight of seat may be neglected. Air resistance may be neglected. Acceleration due to gravity is constant. Child may be modelled as a particle.
  (c) Take into account weight of seat, air resistance, mass of ropes, variation of g, size and shape of the child.

2 $T = \dfrac{mV^2}{a} - mg(2 - 3\cos\theta)$, $H = \dfrac{27a}{16}$

3 $\dfrac{4a}{9}$, $R = 5\frac{2}{3}mg$

4 (a) $m\sqrt{5lg}$ (b) $T = mg$ (c) $\dfrac{l}{3}$

6 $\dfrac{1}{a^2}[u^2 - 2ga(1 - \cos\theta)]$

7 $\dfrac{m}{l}[u^2 - gl(2 - 3\cos\alpha)]$

8 $\cos\theta = \frac{2}{3}\cos\beta$

9 (b) $\dfrac{2mg}{l}(x^2 + lx)$

  (c) $\frac{1}{2}mgl + 2mgx - mv^2$

  (d) $v = \sqrt{\left(\dfrac{3gl}{8}\right)}$

10 (c) $e = \sqrt{\left(\dfrac{2a}{2a+b}\right)}$

11 (b) 6.64 m s^{-1} (c) 4 m

12 (a) $3l = \lambda(0.6 - l)$ (b) $\lambda = 15$
  (c) $60°$ (d) 0.63 m (e) 3.04 m s^{-1}

14 (a) $u = 15.4$ m s^{-1}

15 (a) $v_A = \dfrac{u}{4}(3 - e)$, $v_B = \dfrac{3u}{4}(1 + e)$

  (d) $\lambda = \frac{3}{2}$

17 (a) 2 m s^{-1}, 4 m s^{-1} (b) 0.6 N s

19 (a) $J = 3mu$ (c) $3mu^2$

  (d) $v_B = \dfrac{u(9 - \lambda)}{3(3 + \lambda)}$, $v_c = \dfrac{4u}{\lambda + 3}$

  (f) $\dfrac{8mu}{3}$

20 (a) $10u$, $3u$ (b) $12mu$
21 $\frac{1}{3}V(1 + e)$, $\frac{1}{3}V(1 - 2e)$, $e = \frac{1}{2}$
22 (b) $e = \frac{1}{2}$ (c) $e = \frac{2}{5}$ or $\frac{3}{5}$

23 (b) $\sqrt{(2ga)}$ (c) $\sqrt{5} - 2$ (d) $\dfrac{mg}{\sqrt{5} - 2}$

24 (a) $u = 15.4$
25 (a) 6.75 N s (b) 56.9 m (c) 25.4 m s^{-1}
  (d) No air resistance, no wind, gravitational force constant.

26 $u_A = \dfrac{14u}{5}$, $u_B = \dfrac{3u}{10}$, $I = \dfrac{18mu}{5}$

27 $\frac{1}{2}u(1 - e)$, $\frac{1}{4}u(1 - e^2)$, $\frac{1}{4}u(1 + e)^2$ Three snooker balls in a straight line on a horizontal snooker table with a smooth cloth surface.

29 $\dfrac{5a}{16}$, $\tan\alpha = \dfrac{16}{45}$ **30** $(H+h):(3H-h)$

31 $\theta = 71.5°$ **32** (c) $a, \dfrac{2a}{\pi}$ (d) $78°$

35 (b) $30°$

36 (b) $\dfrac{8a}{9}$ (c) $\frac{14}{15}Mga\sqrt5$ **37** $\dfrac{2a}{\pi}$

38 $\dfrac{10a}{3}$ **39** $\dfrac{1}{\sqrt3}$ **40** $\dfrac{72Mg}{125}$

41 (b) $k=\frac{4}{3}$ (c) $\frac{1}{3}\sqrt{10}\,Mg$

42 $\tan\theta = \frac{3}{4}$

43 (b) $\frac{1}{2}Mg\cot\theta$, $\frac{1}{2}Mg(\tan\theta+1)$

45 (a) $\tan\theta = \frac{1}{5}$
   (b) $1.22\,Mg$, $35°$ to horizontal

46 $\mu = 0.247$ A man whose weight is twice that of the ladder climbing the ladder safely to the top. Ladder modelled by uniform rod. Man modelled by a particle. Wall and floor equally rough. Man can climb to top of ladder if μ is greater than 0.247.

47 $mga(6\cos\theta - 3\sin\theta)$, $\sqrt{205}mg$

49 Ladder may be modelled by a uniform rod. Man may be modelled by a particle. No frictional force at the wall. Model could be improved by taking into account the frictional force at the wall since no surface is perfectly smooth.

50 $x = 2.7$ Model ladder as a uniform rod. Model man as a particle. Assume no frictional force at wall. Model can be refined by taking into account the frictional force at wall.

51 (a) $784\,\text{N}$
   (b) $x = 2\,\text{m}$ Cord light and inextensible. Lamp modelled as particle. Metal rod modelled as a uniform rod.

Examination style paper M2

1 $\sqrt2\pi\,\text{s}$

2 $2\,\text{N\,s}$; Puck, since small, can be modelled by a particle. Ice rink, since there will be little frictional resistance, by smooth plane.

4 (b) $x = 2e^t - 2$

5 (a) $300\,\text{N}$ (b) $78.7°$
   (c) The hammer may be treated as a particle. The arm plus wire is treated as light wire. Wire is regarded as inextensible. Air resistance is neglected. Arm plus wire taken as a straight line.

6 (b) $8\,\text{m}$

7 (b) $\frac{1}{9}(1-2e)(1+e)u$, $\frac{1}{9}(1+e)^2 u$

9 (a) $84.9\,\text{N}$ (c) $m = 7\frac{1}{2}\,\text{kg}$

List of symbols and notation

The following symbols and notation are used in the London modular mathematics examinations:

{ }	the set of
$n(A)$	the number of elements in the set A
$\{x: \ \}$	the set of all x such that
$\in$	is an element of
$\notin$	is not an element of
ϕ	the empty (null) set
$\mathscr{E}$	the universal set
$\cup$	union
$\cap$	intersection
$\subset$	is a subset of
A'	the complement of the set A
PQ	operation Q followed by operation P
$f: A \rightarrow B$	f is a function under which each element of set A has an image in set **B**
$f: x \mapsto y$	f is a function under which x is mapped to y
$f(x)$	the image of x under the function f
f^{-1}	the inverse relation of the function f
fg	the function f of the function g
$\circ\!\!-\!\!-\!\!\circ$	open interval on the number line
$\bullet\!\!-\!\!-\!\!\bullet$	closed interval on the number line
$\mathbb{N}$	the set of positive integers and zero, $\{0, 1, 2, 3, \ldots\}$
$\mathbb{Z}$	the set of integers, $\{0, \pm 1, \pm 2, \pm 3, \ldots\}$
$\mathbb{Z}^+$	the set of positive integers, $\{1, 2, 3, \ldots\}$
$\mathbb{Q}$	the set of rational numbers
$\mathbb{Q}^+$	the set of positive rational numbers, $\{x : x \in \mathbb{Q}, x > 0\}$
$\mathbb{R}$	the set of real numbers
$\mathbb{R}^+$	the set of positive real numbers, $\{x : x \in \mathbb{R}, x > 0\}$
$\mathbb{R}_0^+$	the set of positive real numbers and zero, $\{x : x \in \mathbb{R}, x \geqslant 0\}$
$\mathbb{C}$	the set of complex numbers
$\sqrt{\ }$	the positive square root
$[a, b]$	the interval $\{x : a \leqslant x \leqslant b\}$
$(a, b]$	the interval $\{x : a < x \leqslant b\}$
(a, b)	the interval $\{x : a < x < b\}$

$\lvert x \rvert$	the modulus of $x = \begin{cases} x \text{ for } x \geqslant 0 \\ -x \text{ for } x < 0 \end{cases}, x \in \mathbb{R}$
$\approx$	is approximately equal to
$\mathbf{A}^{-1}$	the inverse of the non-singular matrix A
$\mathbf{A}^{\mathrm{T}}$	the transpose of the matrix A
$\det \mathbf{A}$	the determinant of the square matrix A
$\displaystyle\sum_{r=1}^{n} \mathrm{f}(r)$	$\mathrm{f}(1) + \mathrm{f}(2) + \ldots + \mathrm{f}(n)$
$\displaystyle\prod_{r=1}^{n} \mathrm{f}(r)$	$\mathrm{f}(1)\mathrm{f}(2)\ldots\mathrm{f}(n)$
$\dbinom{n}{r}$	the binomial coefficient $\dfrac{n!}{r!(n-r)!}$ for $n \in \mathbb{Z}^+$ $\dfrac{n(n-1)\ldots(n-r+1)}{r!}$ for $n \in \mathbb{Q}$
$\exp x$	e^x
$\ln x$	the natural logarithm of x, $\log_{\mathrm{e}} x$
$\lg x$	the common logarithm of x, $\log_{10} x$
arcsin	the inverse function of sin with range $[-\pi/2, \pi/2]$
arccos	the inverse function of cos with range $[0, \pi]$
arctan	the inverse function of tan with range $(-\pi/2, \pi/2)$
arsinh	the inverse function of sinh with range $\mathbb{R}$
arcosh	the inverse function of cosh with range $\mathbb{R}_0^+$
artanh	the inverse function of tanh with range $\mathbb{R}$
$\mathrm{f}'(x), \mathrm{f}''(x), \mathrm{f}'''(x)$	the first, second and third derivatives of $\mathrm{f}(x)$ with respect to x
$\mathrm{f}^{(r)}(x)$	the rth derivative of $\mathrm{f}(x)$ with respect to x
$\dot{x}, \ddot{x}, \ldots$	the first, second, $\ldots$ derivatives of x with respect to t
z	a complex number, $z = x + \mathrm{i}y = r(\cos\theta + \mathrm{i}\sin\theta) = r\mathrm{e}^{\mathrm{i}\theta}$
$\operatorname{Re} z$	the real part of z, $\operatorname{Re} z = x = r\cos\theta$
$\operatorname{Im} z$	the imaginary part of z, $\operatorname{Im} z = y = r\sin\theta$
z^*	the conjugate of z, $z^* = x - \mathrm{i}y = r(\cos\theta - \mathrm{i}\sin\theta) = r\mathrm{e}^{-\mathrm{i}\theta}$
$\lvert z \rvert$	the modulus of z, $\lvert z \rvert = \sqrt{(x^2 + y^2)} = r$
$\arg z$	the principal value of the argument of z, $\arg z = \theta, \text{ where } \left.\begin{matrix}\sin\theta = y/r \\ \cos\theta = x/r\end{matrix}\right\} - \pi < \theta \leqslant \pi$
$\mathbf{a}$	the vector $\mathbf{a}$
$\overrightarrow{AB}$	the vector represented in magnitude and direction by the directed line segment AB
$\hat{\mathbf{a}}$	a unit vector in the direction of $\mathbf{a}$
$\mathbf{i}, \mathbf{j}, \mathbf{k}$	unit vectors in the directions of the cartesian coordinate axes
$\lvert \mathbf{a} \rvert$	the magnitude of $\mathbf{a}$
$\lvert \overrightarrow{AB} \rvert$	the magnitude of $\overrightarrow{AB}$
$\mathbf{a}.\mathbf{b}$	the scalar product of $\mathbf{a}$ and $\mathbf{b}$
$\mathbf{a} \times \mathbf{b}$	the vector product of $\mathbf{a}$ and $\mathbf{b}$

A'	the complement of the event A
$P(A)$	probability of the event A
$P(A\|B)$	probability of the event A conditional on the event B
$E(X)$	the mean (expectation, expected value) of the random variable X
X, Y, R, etc.	random variables
x, y, r, etc.	values of the random variables X, Y, R, etc.
$x_1, x_2, \ldots$	observations
$f_1, f_2, \ldots$	frequencies with which the observations $x_1, x_2, \ldots$ occur
$p(x)$	probability function $P(X = x)$ of the discrete random variable X
$p_1, p_2, \ldots$	probabilities of the values $x_1, x_2, \ldots$ of the discrete random variable X
$f(x), g(x), \ldots$	the value of the probability density function of a continuous random variable X
$F(x), G(x), \ldots$	the value of the (cumulative) distribution function $P(X \leqslant x)$ of a continuous random variable X
$\text{Var}(X)$	variance of the random variable X
$B(n, p)$	binomial distribution with parameters n and p
$N(\mu, \sigma^2)$	normal distribution with mean μ and variance σ^2
μ	population mean
σ^2	population variance
σ	population standard deviation
$\bar{x}$	sample mean
s^2	unbiased estimate of population variance from a sample,

$$s^2 = \frac{1}{n-1} \sum (x - \bar{x})^2$$

ϕ	probability density function of the standardised normal variable with distribution $N(0, 1)$
Φ	corresponding cumulative distribution function
α, β	regression coefficients
ρ	product-moment correlation coefficient for a population
r	product-moment correlation coefficient for a sample
$\sim p$	not p
$p \Rightarrow q$	p implies q (if p then q)
$p \Leftrightarrow q$	p implies and is implied by q (p is equivalent to q)

Index